5 STAR SERVICE

Books that make you better

Books that make you better. That make you *be* better, *do* better, *feel* better. Whether you want to upgrade your personal skills or change your job, whether you want to improve your managerial style, become a more powerful communicator, or be stimulated and inspired as you work.

Prentice Hall Business is leading the field with a new breed of skills, careers and development books. Books that are a cut above the mainstream – in topic, content and delivery – with an edge and verve that will make you better, with less effort.

Books that are as sharp and smart as you are.

Prentice Hall Business.
We work harder – so you don't have to.

For more details on products, and to contact us, visit **www.pearsoned.co.uk**

HOW TO DELIVER
EXCEPTIONAL CUSTOMER SERVICE

5 STAR SERVICE

SECOND EDITION

MICHAEL HEPPELL

Prentice Hall Business is an imprint of

Harlow, England • London • New York • Boston • San Francisco • Toronto • Sydney • Singapore • Hong Kong
Tokyo • Seoul • Taipei • New Delhi • Cape Town • Madrid • Mexico City • Amsterdam • Munich • Paris • Milan

PEARSON EDUCATION LIMITED
Edinburgh Gate
Harlow CM20 2JE
Tel: +44 (0)1279 623623
Fax: +44 (0)1279 431059
Website: www.pearsoned.co.uk

First published in Great Britain in 2006
Second edition 2010

ISBN: 978–0–273–73438–3

British Library Cataloguing-in-Publication Data
A catalogue record for this book is available from the British Library

Library of Congress Cataloging-in-Publication Data
Heppell, Michael.
Five star service : how to create magic moments for your customers that get you noticed, remembered, and referred / Michael Heppell. -- 2nd ed.
p. cm.
ISBN 978-0-273-73438-3 (pbk.)
1. Customer services--Management. 2. Consumer satisfaction. I. Title.
HF5415.5.H424 2010
658.8'12--dc22
2010009676

10 9 8 7 6 5 4
14 13 12

Cartoons by Steve Burke, The Design Group

Typeset 11pt Minion by 3
Printed and bound in Great Britain by Henry Ling Limited, at the Dorset Press, Dorchester, DT1 1HD

This book is dedicated to you – the customer

Without you there would be no need for five star service

Contents

Acknowledgements

Thank you!

There are loads of people who give their time and energy to make sure you have a great book in your hands.

First my brilliant publisher, Elie Williams, who tirelessly encourages me and offers carrot and stick in just the right proportions. It's amazing to work with a publisher who focuses more on developing what's right than criticising what's wrong.

Next up it's my writing partner, co-director, best friend and my missus, Christine. Gosh I could go all soppy now but Christine would just cut it out on the first of her many edits!

I have a brilliant team who make sure Michael Heppell is where he should be, doing what he does with an audience in place and much more besides. Laura, Vanessa, Ruth, Alastair and Sheila you are amazing – yes, 'It is a brilliant day at Michael Heppell,' because of you!

I really do think I'm very lucky to have the team at Pearson as my publisher. Writing a book is just the start. Editing, proofing, designing, more proofing, printing, selling, marketing and distribution are a few of the headlines for all the other bits. Thank you for doing this work tirelessly.

Lots of people contributed to this book with stories, ideas and enthusiastic support. Here's a list of some who spring to mind. Jonathan Raggett, everyone at Red Carnation Hotels – they are the best in the world for a reason, Capt. Denny Flanagan, Richard Baker, Peter Williamson, Brian Stanners and his team at Formula One, Linda Eastwood, Terry Laybourne and his team at 21 Hospitality Group, and Steve Burke at The Design Group in Newcastle who created the amazing illustrations.

And, of course, thank you to you, dear reader, for investing your time and money into *5 Star Service*. I hope you enjoy reading it as much as I enjoyed writing it.

Introduction

Welcome to the second edition of *5 Star Service*. Perhaps a good first question to ask would be, 'Why write a second edition?' Well, since the first edition was published in 2006 I'm sure you will agree with me that your customers have become even more demanding. I'm sure you'll also agree that the pressure on you to deliver amazing service has increased and the expectation is that everyone should be better. Me too, so here's how I tackled the challenge of creating a new edition.

Step one was to re-read the original *5 Star Service* manuscript and take out everything that wasn't as important, didn't work or just felt out-dated. After doing this with gusto I managed to delete only two small chapters from the original text (and if you feel you've missed out, I've put them on my website as a free download).

Step two was to continue my research into creating five star service. This was the easy part as I'd never stopped! Every day I'm searching for the best examples of great customer service (for me to follow) and the worst (to avoid!) and, through this constant research, I've added 20 new chapters and dozens of ideas to help you create your own amazing five star service culture.

Step three was to focus on making this book as practical as possible. You, like me, will have read many books, become inspired by the ideas and then ... nothing happens. For this reason I've created a whole new section on how you can implement the ideas from *5 Star Service* with a customer training programme for your organisation.

All in all, the second edition of *5 Star Service* has turned out to be a bumper edition packed with ideas, stories, motivation, tools and techniques to help you become the master of five star service.

You probably love five star service – it makes you feel special, liked or even loved. When you receive it you not only feel good, but you also become forgiving and above all you tell people about it.

And that's the key. People talk about really great service and really poor service but rarely what goes on in between. So if you're thinking 'I don't get any complaints' then that's exactly why you must read this book and test out the ideas immediately. Not getting complaints does not mean you provide great service – it means people aren't complaining *to you*! And when you think about it, that's pretty scary. How many times have you been unhappy with the service you've been given but didn't mention it? Answers on the back of a football pitch please. We are in a massively changing world and one of the biggest changes I've noticed is that customers demand more, from less, and they don't even tell you!

So who are these customers who are demanding this high level of service? There are two main breeds of customer: external – that's everyone who isn't part of your organisation, and internal – that's everyone who is. In a nutshell you can think of your customers as 'anyone who isn't me'. And all of those people deserve five star service.

Why bother with five star service?

Does it sound like a lot of effort to deliver five star service? Maybe you are wondering whether it's really necessary – after all, unless you are a five star hotel, why put in all the extra effort if you see your organisation as having a middle-range market – won't three star service be enough if they are paying three star prices? And what about if you are not in a hotel at all but a mid-sized company that manufactures brake pads – is five star service really necessary for you? What if you work for the Health Service or one of the many other public-sector organisations – is five star service relevant to you? The answer is a resounding YES! In fact, never before has it been more important for you to learn and apply these techniques if you want to achieve targets, get noticed (get promoted), earn more, achieve more and be an altogether better you.

If it costs little or nothing to do, and gives you a huge head start over the competition, why would you *not* want to make your customers happy and win new ones? Could you ever have too many customers? Wouldn't it be a nice problem to have so many happy customers that you needed to expand or raise your prices?

The easy way to five star results

So how do you do it? You'll quickly notice that this book is an easy read – it must be, it's been written by a very simple person. I believe in simple models, not because I think you wouldn't understand more complicated ones but because simple works. During my study for this book I met a researcher who had spent three years looking at a very specific area of customer service and the impact it was having on a certain sector. After three years he had pages of complicated data, wonderful graphs and fabulous flow charts. The challenge was, the people who paid him a fortune to study the problem were not using a single one of his ideas. He knew the answers but couldn't get people to do the 'stuff'.

My thinking is rather different – if you can't pick up an idea, use it immediately and see an impact from it, it won't have made it into the pages that follow. I'd actually go so far as to say that some of the ideas are so simple you'll wonder why I've even bothered to write about them. In response to this I would ask you to observe how many people actually *do* the ideas I write about on a regular, consistent basis. I often hear comments like 'That idea you were talking about, I know that already.' My reply is always the same, 'Great that you know it – do you do it?'

The **secret** isn't in the knowing, it's in the **doing**.

Not every idea in this book is revolutionary. Some you will have thought of before. But if this book just makes you actually take action on the more obvious steps, it will have done its job.

Beware, however, of merely picking out the simple ideas and putting others on the 'too-hard list'. Just smiling, saying 'please' and 'thank you' and remembering people's names won't do very much if your systems don't support a five star service approach. Great design is always great design, great personal service can change like the wind.

How to get the most from this book

You can imagine that writing a book like this, which is designed to appeal to many different people and many different sectors, could be challenging. When I started to put together the ideas I found myself becoming concerned that some of the ideas illustrated by the example of a restaurant might not be suitable for those who work in an office. I spent many hours wondering how to segregate the ideas and make them work for all different sectors, until a friend of mine read the first few chapters and said, 'I can use all of those ideas' – she's a primary school headteacher. The point is, if you're smart (and I'm sure you are) you won't look at an example from a hotel and think that because you don't work in a hotel it's not relevant to you. You'll get creative and ask yourself, 'How does this idea apply to me?' That way of thinking will help you to get the most out of this book.

You'll find some chapters are only a few paragraphs long – don't think the number of words relates to the powerfulness of the ideas. Some of those paragraphs have the very best five star service ideas in them and if you want to get noticed, remembered and referred they are the ones to apply immediately.

Isn't five star service something that only works for a whole team?

Imagine you've read the book, you're applying the ideas but you have a bunch of people around you who think that five star service is just for fancy hotels and first-class flights. Well, it shouldn't really matter that you're the only one who's working in a five star way. You are the most important person in the world, and it's your values that will ultimately make a difference to how well you have lived up to the five star standards. Knowing that you did everything you could to create five star service you'll be able to hold your head up high and know that you did the right thing. You might not always know just how much pleasure you have given to people by your level of service, because just as customers rarely complain, unfortunately they also rarely commend. But you will know you made a difference. And that will also make a difference to your career.

There are lots of ways you can use this book. It's been written so you can just jump in and out and test a few ideas at a time to see which ones work best for you. There are several team exercises where three or four minds will be better than one, but you can get the idea and have a go on your own if it works for you that way. You can read it from cover to cover or just read an idea a day.

What would people say about you?

I know that the vast majority of the stories cited are true because they happened to me. The rest have been checked out as far as is possible and if there's a little poetic licence from those who told me their stories then there's a lesson there too.

As I checked out the stories I realised that some had been exaggerated for either good or bad and either consciously or subconsciously the person in the story was made to be even better or even worse than they had actually been when the original situation occurred.

I bet you've exaggerated a story in your time too.

Imagine if you were the subject of one of these stories, would you like the teller to exaggerate how brilliant you were or how bad you were? **Either way, if you do something, good or bad, that warrants a story, it will be exaggerated.**

What do you want people to exaggerate about when talking about you? Won't it be wonderful when you are being described in such a way that makes you feel so special that you want to get out there and do it again and again?

5 Star Service is designed to help you be the ultimate professional and create magic moments that get you noticed, remembered and referred. Those who do, get more, and those who don't – don't. So read on and remember:

Once you've read it, don't just learn it – **do it!**

1

The Service Star™

Here's a simple way to measure month by month how well you and your organisation are doing in the five star service stakes.

You can do this exercise as an individual or as a team, and it is a great starting point to see just how well you are doing and where you need to improve. For the first couple of times it's important to follow the instructions carefully but after a while you'll be able to do a spot check in just a few minutes and see how well (or how poorly) you are doing.

This simple diagnostic tool obviously works best when you are completely honest. There's no point in giving the scores you hope for as you won't be able to measure your improvement. Look at it as you would a medical – you wouldn't want the doctor to say you were fine if she'd found something wrong that could easily be fixed. It's the same here,

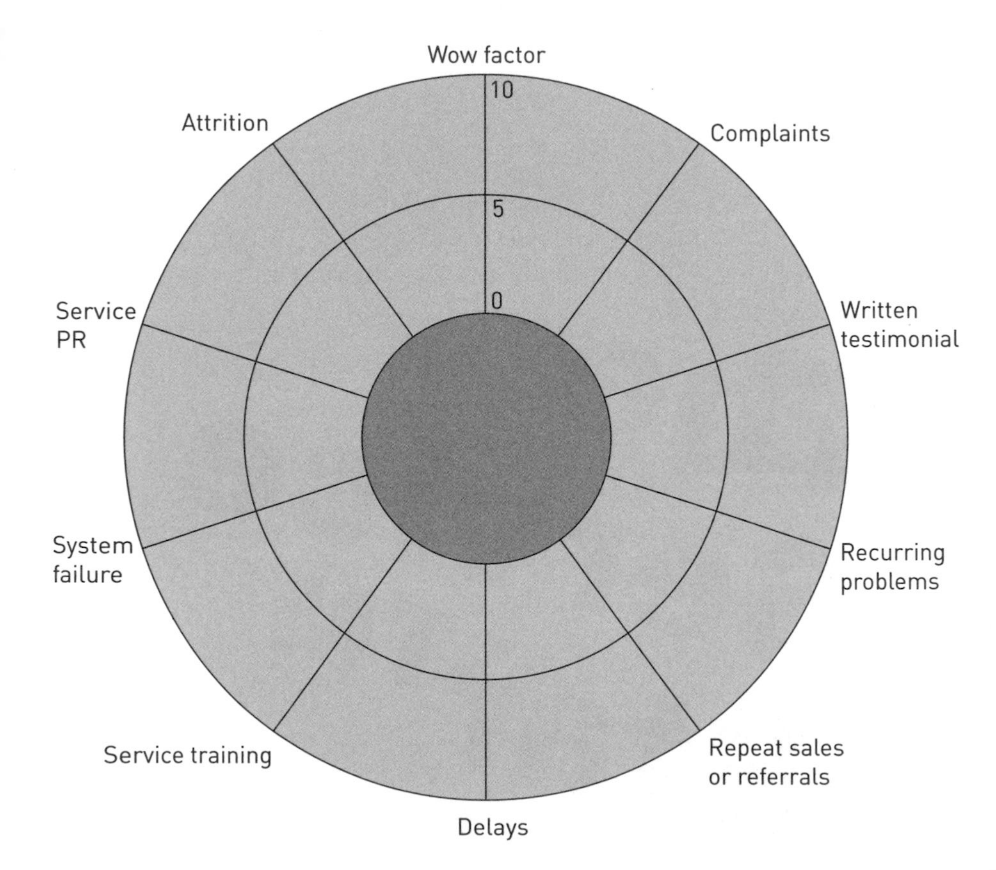

sometimes the biggest challenges are the most exciting area to conquer first – you just need to know.

You'll score yourself out of 10 for each 'spoke' on the Service Star™. Be careful that you read the description carefully though as in some areas you may actually do well by getting a lower score.

You can score this as an individual, for your whole organisation or just a department. It's up to you, but whichever way you choose, be consistent in your approach.

Step one is easy. Just write today's date on the top left corner. If you don't want to write in your book, or want to do this as a team exercise, you can download more Service Stars™ for free at www.michaelheppell.com.

Wow factor

Do you wow your customers? Are you the type of person who is known for the amazing wows they give their customers every day? Are you the one who always seems to find a way to connect and make their day? Do you turn the complaints into compliments and have a reputation for being able to triumph over disaster? If that's you – you're a 10!

Or do you just get on with it? It's not really anyone's job to wow, we'd rather focus on staying afloat and customer service can be someone else's responsibility. If that's you – you're probably a 1 or a 2.

Or you may find you're somewhere in between. Give yourself a mark.

Complaints

Do you get people complaining about you, your department or your company? I'm talking internally and externally here. Do you complain about other people in your organisation and the service they give you? If you feel like there's a lot of complaining going on then that's a *high* mark, an 8, 9 or even a 10 if you don't think it can get any worse!

Or do complaints happen so rarely that it's almost a unique event? Internally and externally? Are you a special breed where it's hard to complain because there's nothing wrong with you, your systems, your service or your customers? If that's the case then you'll score nice and low on this one. Give yourself a mark.

Written testimonial

Is your post bag bursting with praise? Has your notice board been extended just to get all the latest thank-you letters up there? Have you had to buy a new server for the positive emails from your customers and colleagues? If so then you're a 10 and you deserve it.

Or does your notice board have the same 'dog-eared' thank-you letter from 1986 stuck next to the health and safety notice? Are you desperately trying to remember a written thank you you received but you just can't place it? Does your organisation have a culture of pointing it out when you get it wrong and being stony silent when you get it right? If that's you or your organisation, you'll get a low mark here.

Give yourself an honest mark for your written testimonial.

Recurring problems

Do you nip it in the bud as soon as a problem occurs? Are you the type who looks for a challenge even before it happens and makes sure a brilliant system is in place that removes the problem before it rears its ugly head? If you do, then you're going to score nice and low in this area. If that's really you, give yourself a low mark.

Or do you have the same challenges coming up again and again? In fact you may have some come up so often you've named them! If recurring customer service problems are part of the way you do things then you'll probably score a high mark in this recurring problems area. Give yourself an honest mark somewhere between 0 and 10.

Repeat sales or referrals

Cha-ching, there goes the till with yet another sale, this time from a customer who was referred by a colleague or a friend. Ring ring, there's one of your best customers on the phone making another purchase because they wouldn't dream of going anywhere else. Listen carefully ...

someone is talking about you with such reverence that you are definitely going to get Employee of the Month – again. If that sounds like you then I think you can safely give yourself a high mark for repeat sales and referrals.

Or do you have to battle for every new customer? Work hard on building new relationships by convincing people of your merits? If you aren't being referred and recommended or your customers aren't coming back then it's a low mark this time. Give yourself a score.

Delays

Think for a moment about your customer's experience. How long does it take before someone picks up the phone? Do they get the right person first time? How long do they have to wait for information?

If you're speedy and you do what it takes to cut out the queues then you'll find yourself with a nice low mark here. However, if your delays add up to days and your systems are switched to slow, it may be you'll be scoring yourself high in the delays department.

Don't delay – give yourself a mark.

Service training

An easy one this time. How much time, energy and resources do you put into service training? Are you committed to weekly huddles and monthly training mornings? Do you get the training resources you need and take the time to make them work?

Or do you tend to drift in and out of service training now and then? Are you at the bottom of the pile when it comes to customer service training? I know – you're just too busy!

If you're committing time to learning and applying, give yourself a nice high mark. If you just can't seem to fit it in, I'm not asking for your excuses, just mark yourself appropriately low. Or you may be somewhere in between. You decide and give yourself a mark out of 10.

System failure

How good are the systems you have? Do they smooth the way or trip you up? If you have no systems at all then you may think it's difficult to know whether they're failing or not. Trust me, if you don't have a system you've failed and that's equal to a high mark for system failure.

The same applies if your system is just too complicated or ambitious. When the right hand doesn't know what the left is doing it means you have a system failure. When a new person has joined your team and after a few weeks they don't know what to do, or how to get the right information, your system has failed. All these outcomes will give you a high mark for system failure.

Or you may have such great systems that failure isn't even considered. It all works perfectly and when you do have a problem you have a system to pick it up and sort it. That would give you a nice low mark for system failure. Give yourself a mark.

Service PR

What's your reputation like? Are you a Ritz Carlton or a British Rail?* Think about what people might say when they talk about you. Are they praising your amazing customer service or are they using you as the example of how not to do it? Worse still, are they just not talking about you at all?

If you're well known for your five star standing and are independently acknowledged for the wonderful way you work, you can safely give yourself a high mark here. Alternatively, if you have a reputation that you think stinks or if you have no reputation at all to speak of, you'll probably need to give yourself a low mark on this spoke.

* British Rail ceased to exist in November 1997 but is still used as the butt of many poor service jokes.

Attrition

Oh, a nasty one to end on. I'm talking about customers and staff here and it ain't pretty. So are you one of the lucky ones with the same remarkable team who just seem to attract amazing customer-focused talent? Are you able to proudly list your customers from 'way back when' and smugly say you've kept 100 per cent of them? Or are you losing them as fast as you find them?

What about the people in your team? Would the 'former staff' reunion need to be staged in a stadium? Attrition costs, no matter which way you look at it.

So if you have low attrition, give yourself a low mark; if you have high then give yourself a high one; or it could be you land somewhere in the middle.

Finally

Now that you've scored each area, join the dots with nice straight lines. If you're perfect then you should find you have a nice five-point star starting with a 10 'wow factor' running round to 'service PR'.

Or you may have a mish mashy lump in the middle or, worse still, you may end up with a 'reverse star' where your point is at the bottom! If so, don't worry – it's your first time and I'm here to help. There are dozens of ideas on the following pages to transform your lazy lump to a five star Service Star™. And here's the good news. Get your stars sharp and pointy for five months running and those five stars will be the making of you and your amazing customer service.

Bonus Bit

If you would like a free audio guide to accompany you as you complete your Service Star™ you can download one from my website www.michaelheppell.com.

The goal – A perfect Service Star™

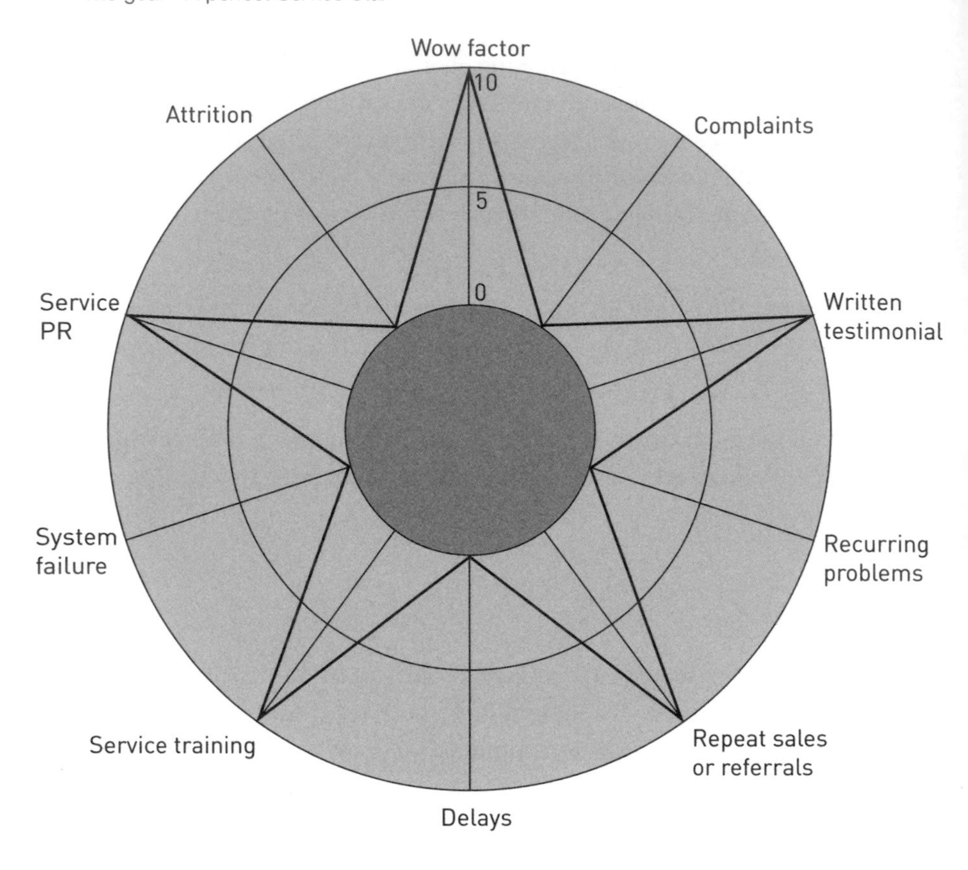

A typical first attempt

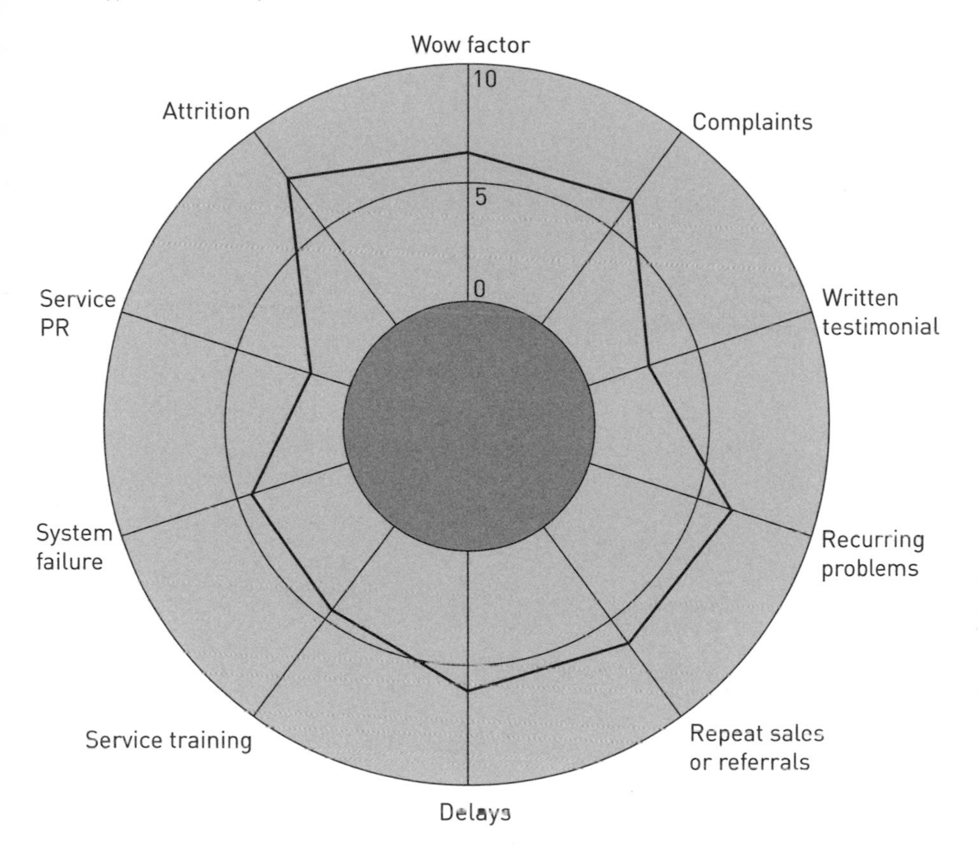

2 Wee Wows

Often we think that the best way to really look after a customer is to wow them. In this book you'll notice many accounts of going beyond the call of duty to wow a customer.

However, there is something we can all do consistently but on a smaller scale. We call them Wee Wows (or mini wows, or son of wow). Wee Wows are the simple ideas that go a long way to making a big difference in how your customers perceive you. One at a time these Wee Wows are great and will help to keep customers 'on side', but do several together and you'll create something very special that your competition will find difficult to replicate.

Here are 20 Wee Wows to get you thinking. Your job is to find 20 more!

1. Never accuse
2. Have a firm handshake
3. Listen carefully
4. Hold the door open
5. Eat slowly
6. Smile
7. Use a person's name
8. Act immediately
9. Make notes
10. Say please and thank you
11. Be positive
12. Check your breath
13. Offer refreshments
14. Be considerate
15. Have pens that work
16. Handwrite 'Yours sincerely'
17. Hand over items the way the customer wants to receive them
18. Establish initial eye contact
19. Tell the truth
20. Do what you say you will do

OK, so they are just wee, tiny little things, but they make a big difference. Next time you are the customer, be aware of how you react when people do 'Wee Wows' and how they make you feel.

Often we can't explain why we feel a certain way about our customer experiences and, guess what? Sometimes your customers can't explain why they feel the way they do about you.

Wee Wows get you focused on doing the little things well.

Here's an added bonus from using Wee Wows. People (friends, family, colleagues, pretty much everyone) will like you a whole lot more. And when people like you, you get better opportunities, you are promoted faster (often you get paid more), you are given the benefit of the doubt and you are followed more readily.

Wee Wows Work!

3

Heads up!

Don't you just love the meerkat? An animal who in the last few years has emerged from relative obscurity to become one of the most widely loved mammals. They look cute, live in communities, eat snakes and weigh less than two pounds.

But did you know that meerkats teach their young the same way humans do? They can be observed demonstrating what to do, encouraging them to have a go, pointing out their mistakes then encouraging them to improve.

And did you know that a meerkat cannot produce body fat, which means they have to find fresh food to eat every day? And, because they are tasty, they have to be very careful when hunting so as not to become the hunted.

What can we learn from the lovable meerkat?

The meerkat's life isn't unlike that of the five star service professional.

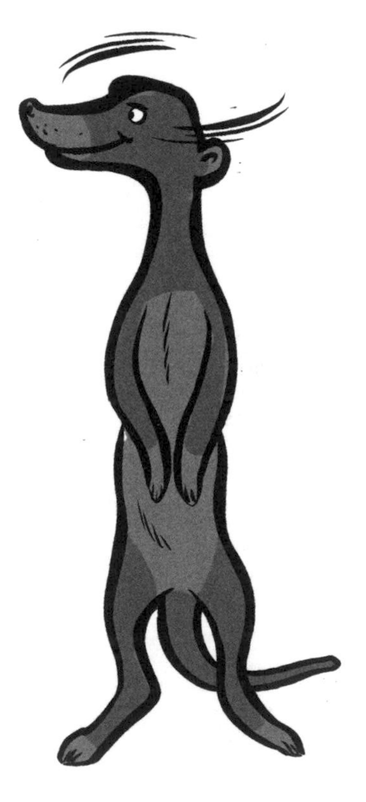

Perhaps the most distinctive member of the meerkat community is the 'lookout'. He is charged with two tasks. Looking out for threats and searching for opportunities. His 'heads up' approach means he doesn't miss a trick.

Here's a big question, which requires a really honest answer. Do you work with your head up or head down?

It's easy to keep your head down and miss the odd opportunity to serve or problem to solve.

Here's how you can use 'heads up' to improve your system. Treat this as a game and make it fun:

Week one

1. Make a set of cards with the words 'Heads Up' printed on them.
2. Give each person in your team a couple of cards and get permission from everyone that you will 'self-police' your actions for the coming week. By that I mean that you will identify times when you think a colleague has missed an opportunity.
   For example, you are busy with a customer, another customer looks like they need help and your colleague seems to be engrossed in something else. You could give your colleague a 'heads up' card and when they say, 'What's that for?' you explain the situation as you saw it.
3. This exercise needs to be playful as the last thing you want is fighting colleagues or an industrial tribunal on your hands. And, if you are the boss you *must* include yourself in the activity. Point out it's for internal customers too.

A week of this activity is usually enough to point out a few challenges, so now it's time to focus on the positive.

Week two

1. Print five cards with a picture of a meerkat on them.
2. In the next seven days, everyone has to give away their meerkat cards to people they observe using a 'heads up' approach.
3. At the end of the week the person who has the most cards wins a prize.

Simple but very effective. And the advantages are:

- You and your colleagues will identify what's wrong (remember to make this light-hearted and fun)
- You'll identify what's right – even better
- People will become more conscious of their own head down or 'heads up' approach.

Think like the meerkat. The meerkat is always looking for **threats** and **opportunities** – you should do the same.

4
Putting on the Ritz

I travel a lot. I was once away from home on business for 130 nights in a year. I didn't stay in the same hotel for longer than two consecutive nights. When you're away so much you become an instant judge of hotels. I know it's not fair to make judgements based on a first impression but hey, we all do it.

Your business might be very different from a hotel, but the two experiences I pick out here perfectly illustrate how the low-cost details make *all* the difference to the customer.

First, the horrendous one.

Glasgow is one of my favourite cities. I lived there for two years and I love the place and the people, so I was really looking forward to going back.

I'd been on the road for 10 days doing lots of speaking and training engagements and arrived at my hotel at 5.30. It was a big city-centre hotel with an impressive foyer. A queue of about 20 people was stretched across it waiting to check in – and only one person was on the front desk. After a few minutes I started to look down the line, hoping another person would join the receptionist. They didn't. After a further 45 minutes it was my turn. I walked to the desk, started to say a nice hello and the receptionist didn't even look up. In fact, as I started to speak she put up her hand to stop me and when she was ready she looked up. Not the best start.

She looked me up and down (I had two cases, a rucksack and computer bag) and she said only two words, 'Checking in?'

'No, I'm here for a colonic,' I found myself saying. Quickly followed by, 'Yes, actually I am.' She had a look on her face which in a nanosecond revealed that colonic hydrotherapy was not high on her list of priorities and she was in no mood to be dealing with the likes of me.

'Name?' That was her next attempt at engaging me. Now I have a surname that can be spelled in several ways, so I always say it, then spell it to help. 'It's Michael Heppell. And Heppell is spelt H. E. P. P. E. L. L.'

She then tapped something into her computer which I know wasn't Heppell, because with great pride she looked up from her keyboard and announced,

'You're not on the system.'

'I should be on the system,' I replied, beginning to regret the joke about the colonic. 'I've got a booking reference and confirmation.'

'Get me the number,' was the affectionate reply.

It took me a further five minutes to put my bags down, find my travel file, locate the booking reference number and read the 28 digits, letters, back slashes and hyphens to her. Then she looked at me, rolled her eyes, and said, 'Oh you meant "Heppell".'

She then asked me to fill in my home address, nationality, etc on the card (passed to me upside down) and to sign next to the two crosses. Then she asked me for a credit card. Have you ever been in that situation where you can't get your cards, wallet or purse out fast enough and it just seems to get wedged as you pull? My cards got stuck at a funny angle in my pocket and the more I tried to pull, the more jammed they became. She repeated in a very impatient tone, 'I said I need a major credit card.' Eventually the cards came free and I handed her some plastic. She literally snatched the card from me, swiped it and slapped it down on the desk with an attitude that screamed,

'We do that just in case you **steal the furniture**.'

Then, as if by magic, she suddenly perked up. A big fake smile appeared (must have done customer service training level 2, this one) and rhymed, 'Would you like a wake-up call or a newspaper? Breakfast is served from seven to nine thirty in the restaurant. Enjoy your stay.'

I jumped into the lift and went up to my room. Dropped my cases and had the usual glance around. Then I was really wowed. I went into the bathroom and noticed that they had folded the end of the toilet roll into a little 'v'. Now I was impressed. I felt like calling home and telling my wife that she had to drop what she was doing and make her way to Glasgow as I'd found a hotel that really knows how to 'wow' a customer as they fold the toilet paper into a little 'v'!

Of course, what I was really doing was wondering why they bother with the traditional 'v' in the loo roll (I know every hotel on the planet does it) when they don't seem to spend as much time working on the very first impression a customer has of the hotel – the staff.

I have had worse experiences, but this was a very expensive hotel with a big brand name and a lot to lose. It's kind of stuck with me and I tell the story a lot.

How could they do better?

Here's my favourite example of how to greet a customer.

In 1997 I visited Singapore and had the privilege of staying in the Ritz Carlton Hotel. I made a promise that one day I would revisit and take my family – it really is a stunning hotel.

In 2004 I was doing some work in Australia and Singapore and it overlapped with the Easter holidays. My son was on his gap year in Oz so we decided to combine the work trip with a holiday. After a long flight to Singapore we arrived at Changi airport where I had arranged for a car to pick us up. The driver was polite and suggested we relax for the short journey to the hotel. Within 30 minutes we pulled up at the front of the hotel – and that's when the magic really began.

It had been seven years since I'd stayed at the hotel and, when the door of the car was opened, the beaming smile of the doorman was followed with a sentence that truly sums up five star service. 'Welcome back to the Ritz Carlton, Mr Heppell.' *Seven years!* Yes, seven years since I'd stayed there and the first words I heard were 'welcome back to the Ritz Carlton'. My wife Christine stepped out of the car after me (remember, this is her first visit) and was greeted with, 'Welcome to the Ritz Carlton, Mrs Heppell. I believe this is your first visit, enjoy your stay.' And as my 11-year-old daughter followed she was greeted with, 'And you must be Miss Sarah, we have a present for you,' and one of the concierge team handed Sarah a small bouquet of flowers.

If it had ended there it would have been great – but it didn't. It just got better and better.

As we walked to reception two members of staff who were walking the opposite way greeted us by name and said how happy they were to have me back and expressed how much they hoped Christine and Sarah would enjoy their stay.

At reception our registration card was waiting for us, printed and completed (with our home address pre-printed on it). It was on the desk, the right way round with a nice pen. All I had to do was sign. *But* – and it's a big but – the receptionist still had to ask me for my credit card. I know that most people reading this book don't work in hotels and many of you will never ask anyone for their credit card, but just learn from this amazing example of five star service from a teenager working as a receptionist. This is what she said after I had signed the registration card.

'Mr and Mrs Heppell, you are going to be staying with us for the next five nights. During your stay you may wish to order room service or perhaps have a drink from the bar. You may wish to purchase something from one of our boutiques or join us for a meal in one of our fine dining restaurants. Perhaps it would be convenient for you to allow me to take a swipe of your credit card so you don't have to worry about carrying money.'

Perfect, perfect, perfect.

What did it cost? The attitude is free. The training costs some initial effort but lasts a lifetime. The system that allows the doorman access to your name after seven years is genius but relies on basic technology. All in all, it's the choice of how you want to treat people.

Does it pay off? I've told this story to audiences all over the world, to FTSE 100 business executives, Fortune 500 leaders. I tell the story on almost every course I teach, certainly at most keynote presentations and pretty much to anyone else who will listen – literally tens of thousands of people from all walks of life.

Many have emailed me to tell me of their experiences and how they went to the Ritz Carlton because I recommended it to them. I never recommend anything formally, I just tell a story. And people are telling stories about you – but rarely TO you: see 'Beware the silent customer' on page 33.

Does your story have a happy ending?

Update

I haven't been to the Ritz Carlton in Singapore for many years, but I have been to the Chesterfield Hotel in London several times. One night my wife and I arrived at the Chesterfield at around 8pm. The receptionist greeted us with a warm, 'Welcome to the Chesterfield, Mr and Mrs Heppell,' before we checked in with minimal fuss and plenty of warmth. When we got to our room I checked my emails and was surprised to find one from my PA who asked if I had made the reservation for the Chesterfield as she hadn't. After a moment of confusion it dawned on me that we weren't actually booked in!

I went back to hotel reception and asked if we had a reservation for that evening. We didn't but here's where the receptionist shone. After her welcome, she checked their system and, even though we didn't have a reservation, she still checked us in, quickly reviewed our history and put us in a room where we had stayed previously and had commented that we'd enjoyed.

She told me that she didn't think I would want to hear the words, 'You don't have a reservation,' as that may have been embarrassing, so simply made us feel good with no fuss and a perfect execution of efficiency and modesty.

5

Complaints – a chance to shine!

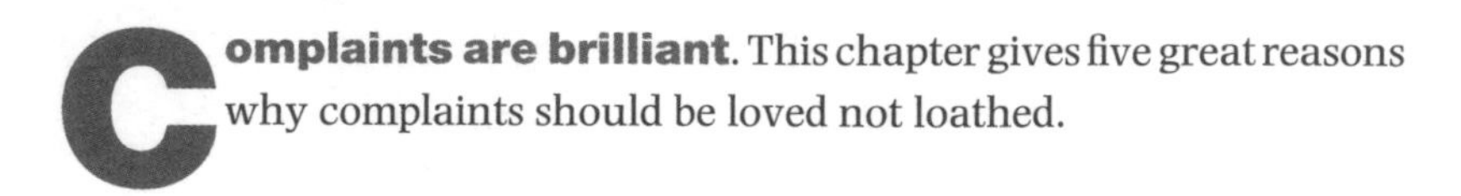

Complaints are brilliant. This chapter gives five great reasons why complaints should be loved not loathed.

You should **love complaints** as they provide an opportunity to **save the day**, **learn from mistakes** and **become better**... but only if you know how.

Reason 1 – Now you know

Ignorance isn't bliss when it comes to customer dissatisfaction. Ignoring an issue doesn't make it go away – it makes it fester and grow, until eventually you feel out of control.

Knowing about a service issue is the first stage to fixing it. Yes it makes you feel slightly sick, especially if you're responsible. Yes you'll wish you hadn't heard it – especially if you care about your customer. And yes you'll be embarrassed – although perhaps you should be. But not for long. Being embarrassed doesn't change things. That's why there's reason number two.

Reason 2 – You get a chance to fix it

Even the best five star service advocates sometimes need a reason to kick some butt to get their point across. Being able to say to colleagues, 'We've had a complaint about this, who'd like to fix it?' makes an issue very immediate. After all, you don't want to receive a second complaint for the same problem do you?

Creating momentum and a desire to fix a problem when you receive a complaint is a good thing. It encourages you (and your team) to take action and overcome procrastination.

The next stage is to close the loop and let your customer know what you have done. Customers love it when you call them, thank them for their complaint, show you care and then explain what actions you have taken to fix it.

Reason 3 – Wake-up call

Now and then we all need a wake-up call. Yes you may be dealing with a 'professional complainer' – you know the type who just complains in an attempt to get free stuff. But if you share my belief that 99 per cent of people are good (see Chapter 11) then you have to take the professional complainer's complaint as seriously as you would any other.

And who knows ... they may be on to something. Imagine having a service culture that works so well that even the professional complainer has nothing to do but enjoy the process?

However, a wake-up call may be just what you need to encourage you to get your finger out and take some proactive service action.

Reason 4 – You learn

Or should I say, I *hope* you learn. Complaints take organisations through a steep learning curve. Then times change, new staff start, ways of working differ and, before you know it, the same problems reoccur.

Here's where a 'Complaints and Solutions Book' is handy. *This is not the same as a public complaints book.* Public complaints books are a crazy idea. Why would you create an environment that announces to customers (and staff), 'We are expecting your complaints'? No, this book is kept in your back office and is used to share with colleagues what happened, how the problem occurred, how you fixed it and, most importantly, what to do so it won't happen again.

Reason 5 – It's better they tell you ...

... than tell their friends. Most people don't complain when something has gone wrong. At least they don't complain to you. Instead they complain to their friends, neighbours, family, in fact anyone with ears. If you're lucky you hear about the complaint second hand and wish they'd said something to you – but they don't so you didn't and they keep on telling their 'poor me' story.

So the next time you're listening to a complaint, it will make the process so much easier when in the back of your mind you're thinking, 'Well at least I know, and I can fix it,' or 'That certainly woke me up, and I can learn from it! And thank goodness they're telling me and not my other customers!'

6

Embracing new technology

Richard Baker was the general manager of Virgin Trains in Wales and north west England. He's also an avid 'Twitterer' and fan of social media. What makes Richard special is the way he has embraced new technology to improve levels of customer service and he's doing it at lightning speed.

Imagine this. You're sat on a Virgin train and you 'tweet' your current feeling about your journey. It may look something like this

> 'On a Virgin train heading to Liverpool. Supposed to be in the quiet coach doing work but there's a really noisy bloke down the carriage on his phone.'

Then imagine a few moments later a Twitter message arrives saying,

> 'Sorry to hear that. Which train are you on?'

You're delighted to get a response, so you message back.

> 'I'm on the 4.30 out of Euston.'

Then the loop is closed when a member of the train crew comes into your carriage and politely asks the noisy bloke to make his call in the 'vestibule area'.

So what happened? Step one is Richard has set up a simple search on Twitter which flags the words 'Virgin' and 'train'. As soon as he sees a comment (good or bad) containing these words he responds whenever he can. On this occasion he knew he could enhance the passenger's journey so he asked him which train he was on. Once he found out this piece of information he called the train and asked a member of the team to take action. They did – job done.

I contacted Richard (via Twitter) and asked him why he used this technology. Here's what he said, 'I use Twitter to talk to our customers. Being responsible for service for a national train company means my customers are all over the country! Twitter allows me to build relationships despite the distance between us. I can also seek out issues and problems people may have had and hopefully resolve them.'

I think the fascinating part of how Richard is using Twitter is not that the technology is making it easier to provide great service – you could argue he has to work harder – but it's how he uses it to be aware, to act and review.

In the very near future the use of this type of technology won't seem exceptional, it will blur into our ways of working just as the telephone has. I'm excited that it won't be unusual to find your suppliers contacting you about a problem that you didn't even tell them about because they used technology to 'scan your output'. 'Your output' being the information you share via social media, customer review web sites, etc. I'd like to think it's a positive type of 'Big Brother'.

I don't want to use this chapter to suggest that you do exactly as Richard Baker has done (although that would be a very good start), but instead to challenge you to think about how you can use new and social media to improve your levels of service.

Here are a few ideas to get your creative juices flowing.

- Create an eCard and send it to your customers. If you don't want to create your own use mine, it's free at www.michaelheppell.com.
- Join Twitter, Facebook and LinkedIn. There's no doubt that this

method of communication is making a massive difference to how we serve our customers. And, like any form of communication, it's up to you to be interested and interesting.

Just promoting how great you are or, worse still, sitting on the sidelines **won't gain you a following**.

- Find a list of consumer websites, find your company and respond to customer complaints, concerns and praises. Don't be defensive if you dropped the ball, be honest and say what you are doing about it.
- Make it easy for your customers to contact you via new media. If you can, make it interactive. To see a great example of this visit the IKEA website and test out 'Ask Anna' on their home page.
- Use new technology to research what other people are doing. If you don't, by the time you've picked up on it, it's too late!
- Write your own blog, and make sure it's relevant to the readers you hope to attract. Allow genuine feedback too. Agree to people leaving comments – good and bad.
- And a little more on that theme – get your customers involved. By allowing your customers to help and debate with each other you'll create an online community. People like to belong to communities.

As I write, all of the ideas mentioned here are reasonably current. However, I am acutely aware that by the very nature of technology, they will soon be out of date. Please don't let this be an excuse for not taking part, go for it with whatever is out there now and you'll soon find yourself giving an amazing five star virtual service too.

7

Beware the silent customer

We are in a changing world. Massive changes are taking place every day in technology, ways of working and tastes. However, I really do believe that the biggest change is that

Customers expect more from less *but they don't tell you*!

Yes, expectations are at an all-time high, we are more demanding than ever, but think about yourself as a customer: do you bother to take time to tell people exactly what you want? My guess is a resounding NO! We just don't have the time.

I was travelling recently with a friend of mine who had a real challenge with his hotel room. When he arrived the room wasn't made up, when he eventually went to his room he found he'd been assigned a twin smoking room. He had booked a double non-smoking one. His shower represented a faint trickle and the heat from the air conditioning was stifling, with no obvious way of turning it down.

We met in reception and he told me all the problems with his room and 'the nightmare' he'd just experienced. Thirty seconds later the duty manager appeared and asked us whether everything was OK. I looked towards my friend and was amazed when he replied, 'Fine, thanks.'

'Fine! What's fine about it?' I asked once the manager had moved on.

'Oh, I can't be bothered. I'll just make sure I don't come back,' he replied with certainty.

So in this situation, as the manager is thinking everybody's happy, the customers are thinking, 'Make sure we don't book here again!'

Even when we left the hotel and we had the customary customer satisfaction form to fill in, my friend still didn't express his dissatisfaction.

So, how do you know what your customers think about you? **Just**

because they are silent doesn't mean they are happy.

What do you do?

1 Create an environment where people feel comfortable giving you feedback – good or bad. You can do this by being open and honest with people right from the start.

2 Ask power questions that are designed to make people share how they really feel. You can do this by asking, 'What can we do better?' Most people will say, 'Erm, nothing.' Because you are now aware of the terror of the silent customer you can say, 'Thank you. If there was one thing, what would it be?' Then you need to listen.

3 When you ask someone to fill in a feedback card, use these words, 'Would you mind completing a feedback card? Please be as honest as you can because we love feedback, it's what makes us better.'

If you do get a customer who complains, remember to thank them. If you think about it, they are probably speaking on behalf of a couple of dozen other customers who thought it but didn't say it. Then use one of the 50 or so ideas in this book designed to make sure you win them over and keep them for a lifetime of loyalty.

Bonus Bit

Make a list of everything you think your customers want, then the next time you meet with them (or a representation of them) show them the list and ask whether you've got it right and then ask whether they would like to add anything else.

You may think this is a bit cheesy and you may be right, but think what you would do if one of your suppliers did that for you. Imagine how you would feel if the next time you went to your dentist a friendly member of staff took five minutes to ask you to look at a list of potential improvements to service and offered you an opportunity to add to the list. Then imagine how it would feel if the next time you went there they'd applied some of your ideas.

8
RADAR
thinking™

What do you do if you have a **common problem** that comes up time after time?

You know the customer isn't going to be happy but you can't do anything about it. It really is outside your control.

Radar is a wonderful invention. It prevents catastrophes every day. It doesn't move the hazard, it just tells you where it is and most importantly gives you time to avoid it. A ship can sail the same route week after week. If the crew got upset and complained about the hazard and spent years trying to move it, it would never go anywhere and the company that runs it would quickly go bust.

So RADAR thinking™ takes situations where you previously thought you had a nightmare and turns them into customer service dreams.

Use RADAR as an acronym:

Realise

Assess

Decide

Act

Review

Realise – that you actually have a problem. Get your head out of the sand and take a look at the regular challenges your customers face. Is it delivery times? Sub-contractors? Costs? Customer stupidity?

Once you have a list . . .

Assess – have a really good brainstorm and throw lots of ideas into the pot. If you really are going to have a powerful brainstorm, make sure you use the three golden rules:

1. Write every idea down.
2. Keep it positive (ban anyone who says we've tried that before and it didn't work).
3. Allow everyone an opportunity to participate.

Once you have assessed all your ideas …

Decide – which ideas you are going to use and which ones you are going to park (or bin) for another day. Take the very best ideas and make sure they complement the real problem (go back to 'Realise' and check).

Once you have one or two that you are committed to …

Act – decisions without actions are pretty worthless. Have you heard the story of the four frogs sitting on the log? One decided to jump off. How many frogs were left? The answer is four because the frog only *decided* to jump off, he didn't actually do it! Ensure you put your ideas into practice swiftly and ensure that everyone knows what you are doing.

Once you have the ideas in place …

Review – take time to fine-tune or, if needs be, ditch an idea. Forget about that nonsense of 'get it right the first time' – go for getting it right by doing it. There aren't many ideas that changed the world by needing to be right the first time. Author's note: if you are a pilot reading this, the 'get it right the first time' idea *does* apply to you!

Once understood, you can use the RADAR technique as part of your weekly meetings or monthly staff training. The more people who are involved with the ownership of the ideas, the better they work.

Next, I'm going to give you three examples of RADAR thinking™ at its best.

9

RADAR thinking™ at work

Here are three examples of RADAR thinking™ being used to turn a regular challenge into a wonderful customer service experience.

1 I guess no customer service book would be complete without an example from Disney. Epcot in Orlando, Florida, is Disney's biggest park and as you would expect has a very big car park. If you have ever been (and driven there) you'll know the scale is huge – the Epcot car park alone can take up to 20,000 cars.

Logistically to get everyone from their cars to the park at the start of the day or from the park to their cars at the end of the day is a huge task so it doesn't help when a weary family gets to the end of a long day and tells a Disney cast member (their name for staff) that they can't remember where they've parked their car. This would be a nuisance if it happened to a dozen or so families but on busy days up to 500 people could have lost their cars.

Conventional thinking would be to suggest to the customer how utterly thick they have been and suggest they wait until everyone else has gone home. Then one of the remaining cars in the 30-acre car park would be theirs – and you hope they've learned a valuable lesson.

It was suggested at one point that Disney could have a lost car waiting area where families could shelter, watch cartoons and purchase snacks. Then when the car park cleared, a special 'Find your car' train would leave with everyone frantically pressing their remote key fobs until some lights flashed. Good fun and a nice idea but at the end of a long day what do you want? That's right – you just want your car back.

Here's how they fixed the problem, allowing over 90 per cent of people to find their car in less than five minutes.

When you are collected from the car park, a road train takes you from your row in the car park to the entrance. The train driver asks you to remember the name of your car park (all named after Disney characters) and the row you have parked in. He even gets you to say it out loud. But most people are so excited (and exhausted) by the end of the day that a car park name and row number are just a distant memory.

So (and here's the Disney magic) the driver who picks you up also writes the exact time he or she arrived, the name of the car park and the row number. Then should you lose your car the cast member assigned to help you locate it will ask, 'Do you remember what time you arrived?' In most cases people remember the time they arrived because they are: a) proud of the fact they got there early, b) rushing because the gates have or are about to open, c) thinking about how much of their day has been lost because they are arriving late (Disney tickets aren't cheap!). Then they look down the list for your arrival time and reveal that you were parked in 'Pluto 15' and a wave of familiarity washes over you.

This simple yet effective system **thrills customers**, frees up **valuable** staff **time** and can be **duplicated effortlessly**.

2 Olympian Furniture in Edinburgh and Glasgow is an amazing company. Not only does it provide great furniture but the staff look after you like you are the only person in the world. It's a pleasure to shop there and they love to have you even if you don't buy anything. However, they have a problem.

Most of the furniture is hand-made and can take up to 12 weeks from order to delivery. Excited customers would fall in love with a beautiful piece of furniture and proudly place an order. They would be told, 'It may take up to 12 weeks for delivery'. As you can imagine, once ordered, when do they want it? That's right – *now* or sooner if possible. The brilliant staff at Olympian used to say, 'It will be worth the wait.' Very true, but customers would still be thinking, 'I want it now.'

So after learning about RADAR they realised that the problem wasn't going to go away unless they started to stock pre-packed lower-quality goods and that was never going to happen.

They **realised** they had a problem with the waiting time.

They **assessed** what they could do and involved every member of the company, including delivery and back-office staff.

They **decided** to turn this wait into a feature by actually building it into their sales presentation.

They took **action** and wrote an overview for all staff to share with customers explaining that to have a piece of furniture of this quality takes longer. They would explain what would happen after their order was placed, how the leather is dyed in Italy or how the wood is polished using a special process that can't be hurried. They would explain that the furniture is delivered complete and must be robust enough for travel and how the delivery team will bring the furniture to them only when it has had a final inspection.

While **reviewing** the strategy they have continued to improve on this technique, showing customers why it's more than 'worth the wait' and it's part of the quality of the Olympian experience.

3 My friend Nicola called me and said, 'Dyson vacuum cleaners use RADAR.' I've never worked with them so I'm sure they don't call their thinking RADAR, but their idea was simple and class.

Nicola's Dyson stopped working. After three years of trouble-free vacuuming it just stopped. The first bit of RADAR thinking™ was obvious when she found the helpline number on the cleaner (because we don't keep manuals). Second was when she called on a Saturday night there was someone there because 'people use Dysons at all times, so we have to be here at all times', advised the customer service agent. A quick diagnostic predicted a burnt-out motor. An engineer would be sent out and it would be a maximum of £50 no matter what the problem was.

The engineer arrived, on time, looking smart, and sure enough the problem was the motor, which he replaced. And he changed the cable because that looked a little worn. And fitted a new plug because the other one was cracked. And replaced the filter and gave the whole machine a smarten up (all for the agreed £50).

Then RADAR really kicked in as the engineer asked Nicola whether she knew how to wash the filter or that she was supposed to do it every six months. No and no came the reply. It turns out that the number one reason why a motor burns out is because users don't clean the filter and you really should do it every six months. So he showed Nicola how to take it out, how to wash it and how to replace it.

Then he asked whether she had a PC and took her to the Dyson website and registered her to receive a six-monthly email reminder to clean the filter, because we could be better at remembering stuff like that.

One other thing, while she was on the site she was offered 10 per cent off attachments and ended up buying another £60 worth of products!

Send your ideas on how you have used RADAR thinking™ to info@michaelheppell.com and, who knows, you could be in the next edition of *5 Star Service*.

10

The emotional bank account

I'm sure you have a bank account and when it's in credit you feel good and when it's overdrawn it's a little more of a concern. You also have an emotional bank account which works in a similar way. When you have received lots of deposits you feel great but when you have experienced too many emotional withdrawals it doesn't feel so good.

Your customers have an emotional bank account too. And when you've made lots of deposits they feel good, remember you (for the right reasons) and recommend you. When you go overdrawn you won't be top of their Christmas card list and you certainly won't have their loyalty.

The challenge is that often we attempt to provide good service to 'get us out of debt'. By then it's too late. The secret is to create lots of small (or large) deposits first, so that when you have to make a withdrawal you won't go into the red.

Here's an example of how an emotional bank account can work for you as a traveller flying on a discounted airline:

You book a cheap flight and save money	**Deposit**
After a short queue the check-in assistant is pleasant	**Deposit**
You're told your luggage is 2 kilos over and you have to pay extra	**Withdrawl**
Security has short queues and you don't get stopped	**Deposit**
The flight is on time	**Big deposit**
Your boarding number is called and the scene is similar to that of a Spanish bull run!	**Withdrawal**
You board the flight and receive a polite, sincere welcome	**Deposit**
Once on the plane you're told that there is a problem with some equipment and you will have to sit on the plane until an engineer can check it out – it takes the next two hours	**Withdrawal**

How the staff handled the first part of you flight experience will make a massive difference to how well you take the news that you have to sit on the tarmac for the next two hours. We've probably all met customers who will do anything to defend the goodwill of an organisation. Why? Generally because they have had so many deposits made into their emotional bank account by that organisation that they will do anything to defend it.

You've probably done it yourself. Have you ever been in a restaurant and had an absolutely amazing meal so you went back again and again? Then one night you went for a meal and the service wasn't up to scratch, the food wasn't to the usual high standard and the evening could be viewed as a bit of a disappointment? I'd gamble you started to defend the restaurant, especially if you had other people with you. 'It isn't normally like this, maybe they've got some staffing problems,' or 'I hope everything is OK, I noticed that Julian the head waiter isn't in tonight. Next time we'll book when he's here.'

That night the restaurant took some withdrawals from your emotional bank account but because it had made so many deposits beforehand you will go back. In fact, if anybody asks you about the restaurant and whether it's a good place to eat you will completely blank the bad experience from your mind and tell them about previous meals and the fantastic service you get from your friend Julian.

However, if you went back another two or three times and continued to receive the same bad service, low-grade food and poor value for money, it wouldn't be long before it would have taken too many withdrawals and had a negative impact on your emotional account. By then it would be too late. Even if Julian was to come and knock on your door begging for you to come back, you probably wouldn't.

Wee Wow

You can never put too many deposits into someone's emotional bank account. Who knows, one day you might need to make a huge withdrawal.

Think about your customers, internal and external. How are their emotional bank balances? Time to make a deposit?

Actions

Here are five things that will instantly create deposits in emotional bank accounts.

1. Look happy. Facial expression is one of the easiest but most effective ways to make a deposit or a withdrawal from somebody's emotional bank account.
2. Do what you say you're going to do. It's one of the top three referability habits and not enough people do it.
3. Never blame the customer for anything. If the customer knows they are wrong and you resist the urge to blame them, you'll get double deposits in the emotional bank account.
4. Listen carefully. If you really listen and replay what a customer has said, you will score some rapid deposits. Why? Because most people don't listen and most people want to be heard.
5. Be sincere.

11

99 per cent of people are good ...

... so please don't treat me like the other 1 per cent.

Here's the deal. I'm shopping. Not my favourite pastime, in fact I'm one of those people who loves the idea of getting all my clothes for the year in one fell swoop. Nothing pleases me more than to find an item of clothing that fits well and looks good, then ordering exactly the same in black, blue, brown, grey and six other colours in between.

On a recent excursion to a very well-known store I had a bundle of stuff that I wanted to try on and I arrived at the changing rooms. The ever-so-efficient assistant announced, 'You can only take a maximum five items into the changing rooms.' Of course I had four times that as I was in shopping mode and stocking up for the winter. Plus I'd be doing anything to avoid coming back. So I had to ask, 'Why?'

'It's a security measure,' she claimed.

'I'm sorry, but I don't feel any more secure by only taking five items into the changing rooms,' was my attempt at a humorous retort.

'No, sir, it's security for the store – to reduce theft.'

At this point she had no idea that by association my subconscious mind had asked, 'Who are you calling a thief?' My conscious mind said, 'I hope she doesn't think I look like a thief!' and my mouth blurted out, 'OK, I'll just take the five at a time.'

But the **damage was done**!

I couldn't try the clothes on in the same way. I didn't feel comfortable in the changing room as I kept looking in the mirror thinking I should be trying on a ski mask or pushing a pair of tights over my head.

You guessed it, I didn't buy a thing. I didn't get past the first five items before waiting until she was distracted and running out of the door. I have an association with the shop which doesn't really float my boat and according to the news the company is currently 'struggling in an ever more competitive high street'.

Simple message. I hate thieves. I despise the fact that I work hard for my money, pay tax on my earnings, pay tax again when I buy a product, queue at the counter and do everything by the book while other people are just taking stuff without paying. I'm sure you feel the same.

So **don't** make me **feel** like you have the slightest doubt in your mind that **I could be one of them**!

I want to be trusted as a customer. Trust is one of our deepest emotional values. When I'm trusted I feel good – when I feel good I *buy stuff*, when I'm buying stuff you can bet other people are buying stuff too.

Here are some examples of other organisations that trust first:

Self-service checkouts – I know they are cheaper to run but if you have ever used one you'll know you scan everything (usually very carefully) to make sure you get it right and pay for everything.

Hotel 41 in London where the guest bar is an honesty bar. You just help yourself, write it down in the book and pay the next day.

The person in the shop who, when they see you are a few coins short, asks you to 'just drop it in the next time you're passing'. I bet you have made a point of paying off a tiny amount because someone trusted you.

So what if you work in a place where 'company policy' dictates that 'we can't allow more than five items in the changing room'? How could the lady at the start of this chapter have handled the situation in a five star way?

Just imagine if, as I walked towards her, she had stepped out from behind the desk and said, 'Here, let me help you with those,' then taken some of the items from me. Then if she had added, 'You've got some great clothes here, let me find a changing room for you. OK, sir, our rooms aren't very big so can I suggest you take four or five items at a time and I'll look after the rest. If you want to change anything just give me a shout and I'll bring it through and swap it for you.'

I'd buy the lot!

I don't want to be treated like the 1 per cent. I appreciate in retail it's a challenge that people steal your stuff but if you don't trust me I'm going to be taken down an emotional peg or two and that could make me unhappy. Unhappy people don't buy as much. Unhappy people tell their friends and anyone else who cares to listen about their problems (you could be one of them). Unhappy people really don't buy as much. Unhappy people make your store look gloomy. Oh yes, and in case you didn't pick it up,

Unhappy people buy less, a lot less!

So the next time I'm buying something from you, 'Trust me – I'm a customer!'

Actions

There are three questions to ask yourself:

1. Do you ever say 'it's company policy'?
2. What could you say instead?
3. When do you 'accidentally' treat people badly?

12

The top three referability habits

A survey was carried out among 1,000 customers which asked them what it would take for them to refer an organisation to their friends or family. One question specifically asked what an individual had to do. The top three answers were:

1 Always **do** what you **say** you are **going to do**.

2 Be on **time**.

3 Always say **please** and **thank you**.

I'm not surprised by the first one. It makes perfect sense that if you are going to promise that something will happen, it's up to you to make sure you deliver the cookie. I remember an old boss of mine once saying, 'We have to under-promise and over-deliver.'

Those days are gone – now you just have to *promise and over-deliver*.

In simple terms, *do what you say you're going to do*. People are reluctant to take responsibility as there are a million good reasons why 'what should have happened didn't'. We blame suppliers, we blame technology, we blame colleagues, we blame outside forces and internal politics, but at the end of the day if we're going to create five star service we have to take the responsibility to get things done on the chin. That's why if you bought this book for yourself or if you were given it by somebody else, I would ask you to think, 'What can I get out of the book for me and how can I make a difference to my organisation?' By doing this you'll provide five star service on a consistent and regular basis.

'Be on time' reminds us that if we say we're going to do something by a certain time, we should do it.

Often we think it's a good excuse not to be on time because we are so busy. Actually it's **because people are so busy** that it's more important than ever to **be on time**.

My friend Jeremy Taylor runs his life on what he calls 'Jes time'. He's a busy guy running a very successful business, but he's always on time for every appointment and every meeting. He understands that roads are busier than ever and most things take longer than planned. If he plans his day well he gets masses done, then gets home in time for his wonderful family.

He works it like this. If he has an appointment at 10 o'clock he plans to be there by 9 o'clock. He would rather sit outside for 20 minutes reading his notes and walk in 10 minutes early than arrive 10 minutes late making apologies and moaning about 'the state of the roads', 'the lack of parking' or 'the overrun of his last meeting'. Jeremy gets lots of referrals.

The third referability habit is more puzzling because it has a very specific point which came out of the survey. Not just to be polite but to actually say your Ps and Qs. When I read it I started to take a mental note of how often I remember to say 'please' and 'thank you' and I asked my team and family to point out when I didn't. I was shocked. I would consider myself to be an incredibly polite person but it's only when you stop and look at your actions and omissions that you really realise how many times you forget to say those simple words.

Here's a challenge. For the next 48 hours, when you engage with anyone be conscious of how many times you do or don't say 'please' and 'thank you'. You may wish to do what I did and ask some of your key people (friends, family, work colleagues) to point out when you forget.

Often it's the small things that make the big difference – see the chapter on *Wee Wows* for more.

Actions

Promise *and* deliver. Take responsibility (the ability to respond) seriously. If you say it's going to be done, make sure it happens.

Work on 'Jes time' and plan to get there early. You can do lots with the extra time you have if you should get there early, or if you are running late it's a lovely feeling to have half an hour or so to fall back on.

Ask you friends and colleagues to (politely) point out when you don't say 'please' or 'thank you'. Make sure you say it sincerely.

13

Ring the bell

Here's a simple but effective idea to promote five star thinking and working in your organisation. And all it needs is a desktop bell. The basic idea is simply to encourage each other to ring the bell every time you have carried out a good customer service act or when you have something to celebrate.

I've used this idea for years and in some organisations where we introduced it, it's become a major feature of their ways of working. It developed because I wanted people to share their successes.

I was working with a company and discovered that a salesperson had just completed a £10 million deal that day. He arrived back in the office, sat at his PC and sent an email to his boss, who the next morning sent a reply saying 'well done'. He was to be congratulated at an event that was taking place six weeks later.

So I started to think. We've all been to similar events. They are great fun and very motivational, but what about the other 360-plus days of the year? How do we celebrate success then? I'd had a bell on my desk for years and would use it to create a 'ding' for my team, but there was nothing formal behind it. The following week I was back with my client and I gave them my bell. We set some rules for what you had to do in order to 'ring the bell'. *Most people could qualify to ring the bell at least once per day.*

Then something really brilliant happened. The first person to ring the bell sheepishly walked up to the desk it was on and gave it a ting. Three or four of her colleagues gave a little cheer, then she was asked the question (which really makes this system work), '*What have you done?*' She then went on to tell a 15-second story of how she prevented a customer from switching to a competitor and saved the company a few thousand pounds.

She was talking about **best practice**.

She was **getting recognition**.

She was creating a **feel-good** atmosphere in the office.

So picture the scene. You are part of a team of 15 people working in an open-plan office and perched right in the middle of all the desks is your celebration bell. Then:

- Decide what will be the criteria for ringing the bell.
- Make sure everyone celebrates when someone rings the bell – even just a little 'wahoo' is better than a stony silence.
- Keep reinventing the idea and make it fresh.

There's one other secret bonus of using the bell technique to celebrate and promote success. What would happen if you hadn't rung the bell for a few days and all your colleagues had? Wouldn't it drive you on to want to ring it?

It really does spur people on to ring the bell on a regular basis – every 'ting' is a successful thing!

Actions

Five-step guide to ringing the bell:

1 Buy a bell (or a bunch of them). Go to www.michaelheppell.com if you can't find one.
2 Place it in a prominent position in the room.
3 Decide on the rules for what you have to do to ring the bell (let people break the rules if they have something to celebrate!).
4 Encourage people to ring the bell by asking, 'Have you rung the bell for that?' whenever you hear about a success. It takes a little while to create new habits.
5 Give people the freedom to tell their story of best practice and record as many of the exceptional ones as you can in your five star journal (see page 234).

Wee Wow

I have one client who bought a gong to bash for anyone who really went the extra mile. They put it right next to the accounts department (a very challenging part of the company) so they would know exactly who was bringing in the money.

The customer is always right – not!

There's an old expression that says 'the customer is always right'. Is it true? Of course not. Customers get it wrong all the time and sometimes that can be to the detriment of you and your organisation. On those occasions you need to know how to react.

When my son was at university, during weekends and holidays he worked for a well-known retailer that offered a fantastic 16-day money-back deal if you are not completely satisfied with the products. We had many conversations about customers bringing things back outside the 16-day policy and asking for a refund. There's nothing wrong with the product, they just changed their minds and left it a little too late.

Would it be five star service just to turn a blind eye to the advertised 16-day policy and give customers their money back on the 17th, 18th or 19th days? I don't think so. This organisation has a very clear method of trading and does it very successfully. However, the way in which you tell a customer that they can't have the money back in this particular instance is very important.

The **last thing they want to hear** are words like 'It's company policy' or 'I don't make the rules'.

But at the same time they do need to be told that there is a rule in place. Sometimes there are customers who will exploit the situation given the opportunity to do so, and they will continue to take more and more. You've heard the adage 'Give them a finger and they'll want your hand. Give them your hand and they'll want your arm. Give them your arm and you'll have nothing left.'

So how do you let customers know that you're not a 'soft touch'

but at the same time make them feel as though they have had five star service?

I believe it's about communicating in a clear, empathetic way with certainty and belief in your voice. Did you have a teacher at school who scared the living daylights out of you and if they told you to have your homework in by Wednesday morning, you did it mainly out of fear? You probably didn't like the teacher but you did as they said. Did you also have a teacher who would ask you to have your homework completed on time, but if you didn't it wouldn't really matter? I bet you liked that teacher! And I bet you had one or two teachers (a rare breed) who would ask you to have your homework completed by Wednesday morning, you did it, *and* you liked the teacher.

What qualities did they have that made you do as you were told and still they remained likeable? My guess is they believed in what they were doing and this came across in their manner. So if you do have to deliver news to a customer that they might not want to hear, make sure you believe in what you are saying first, then do it in a polite but firm way, and if they still don't like what you say then skim to 'It's your best friend – the awkward customer' on page 97.

There's a difference between confidence and arrogance that needs to be explored here. When you are confident you have a certain demeanour that in most cases is seen as attractive.

People like confident people and like to be guided and led by them.

Beware arrogance is only a few degrees away from confidence but instantly turns people off.

The challenge is when you are 'trying' to be confident you can easily come across as arrogant. The key word here is 'trying'. Confidence needs to be a natural part of what you do rather than forced. So how do you do it?

Know your stuff. You get this in two ways: studying it and doing it. When you become an expert in your ways of working, your products, your methods and your systems, you automatically become more confident. But the big one comes when you do it. *Every time you use a tool or technique it becomes easier, the easier it becomes the more confident you become, the more confident you become, the more in control you feel.*

Sometimes it's very difficult to do all of those things at once but there is a short cut, and the next section features three of the most powerful words I have ever learned which will show you how. When you use these three words in the right way you can defuse almost any situation and align your customers to your way of thinking. The three magic words are:

Feel

Felt

Found

Read in the next chapter how you can use those magic words in many difficult situations, including the one we started this chapter with.

15

Feel, felt, found

'**Feel, felt, found**.' Three magic words and the best 'aligning' technique I've ever heard. They work so well and are easy to remember. This technique works very well with situations that are outside your control, you can stay strong and you'll get people coming to your way of thinking. Here's why.

Imagine the situation. You have to listen to a customer who wishes to complain about something that you cannot do anything about. You could easily say, 'I'm sorry, but company policy dictates that I can't do anything about this' – if there was ever a way to upset a customer it has to be starting a sentence with 'company policy dictates'.

This is why 'feel, felt, found' works so well.

Let's start with 'feel'. Who wouldn't like somebody who appreciates how another person feels? It's an instant way to align with a customer and at the same time show empathy. Here is how you could use the word in a sentence, 'I'm sorry you feel that way' or 'I can understand how you might feel.'

Now let's take a look at the word 'felt'. Felt is a powerful word because it does two things. It appears that you are viewing the situation from an historic perspective. It shows that you have experience of this situation. That's where the last word, 'found', comes in.

'Found' makes you the expert.

When people have 'findings' it usually means they have done a huge amount of work and research, and that research has come up with a fact, solution or answer. When you share your findings with your customer, you demonstrate your expert knowledge and you also empathise with their emotions. Lovely, isn't it?

'Feel, felt, found' has a remarkable way of breaking down barriers and aligning customers to your way of thinking. Here are a couple of examples of how it can work in practice.

Customer: 'You mean to tell me it's going to take two weeks for it to be delivered? I think that's ridiculous.'

You: 'I understand how you might feel. Two weeks does seem like a long time. And before I started working here I probably would have felt the same, but do you know what I found? Because all of our products are ordered directly from the manufacturer we always ensure that we have the latest model to give to our customers. I also found that carrying a large amount of stock doesn't give us the opportunity to offer you the lowest price on the very latest models. We found most customers would rather have the latest model at the very best price.'

or

Customer: 'This is the third time I've had to call you with the same problem. And every time I get through to a different person and have to explain it all over again.'

You: 'I understand that you must feel frustrated with this. If I had explained the problem three times, I'm sure I would have felt the same. I've found that if I hear the details directly from our customer I am able to understand the issues more clearly and resolve the problem more rapidly, rather than reading some notes from a computer screen.'

And finally, from the previous section, what do you do when a customer is outside an already very generous returns policy and there's nothing wrong with the product?

Customer: 'I'd like a refund on this.'

You: 'I'm very sorry, but we can't give a refund outside of our 16-day satisfaction guarantee.'

Customer: 'That's terrible, it's only day 18. What difference does a couple of days make?'

You: 'I understand that you may feel unhappy about that and I have had other customers who have felt the same. But we've found that our satisfaction guarantee is one of the most generous in the high street and we've also found that the vast majority of people are delighted with it and can easily get back to the store in 16 days.'

Final thought:

If you're not sure this will work for you then I understand why you might feel that way. In fact before I actually tried it, I felt the same, but once I tested out the idea I found it worked brilliantly!

16

What's in a smile?

From just a few weeks after birth, we demonstrate that we feel happy by smiling. All through our lives and until our dying day we use the same process. People are even heard to comment that they'd like to die with a smile on their face.

So take a look around. If smiling is so important, why aren't more people doing it? The next time you walk down the street check out how many people are smiling. You won't be surprised to find that it's often a small minority who find something to smile about. Even when there's good news, people often don't smile.

In these days of high competition, where price and speed crash headlong with convenience and perceptions painted by the media, one of the purest things we can do is to offer a sincere smile.

A smile is timeless.

A true smile, one that resonates from the heart, can lift a person and completely change the way they feel. A smile is directly linked to your internal physiology. In other words, when you smile, you connect with your entire nervous system and your whole body reacts in a different way from when you're not smiling.

You have probably read or heard that it takes 17 muscles to smile and 43 to frown (or similar). The truth is it actually takes 12 muscles to smile and just 11 to frown. Here's the breakdown for you facial muscle freaks out there.

Muscles involved in a 'zygomatic' (genuine) smile: zygomaticus major and minor, which is four. Orbicularis oculi – that's two. Levator labii superioris – two again. Levator anguli oris, which also helps lift the angle of the mouth – two more. And your risorius which pulls the corners of your mouth to the side. Again two of these are needed or your smile is one-sided so that gives a total of 12.

And here are the muscles needed to frown: orbicularis oculi (again), that's two. Platysma – another two. Corrugator supercilii (bilateral) and procerus (unilateral) gives you a furrowed brow so that's three.

Orbicularis oris, that's another unilateral so there's just one. Mentalis is another single so again just one and finally good old depressor anguli oris which pulls down the corner of your mouth, two there, giving a frowning total of 11.

But it's only one muscle less, so resist the temptation!

Anatomy aside, don't people prefer smiles? As customer service professionals there has never been a more important time to learn how to smile.

In the movie *Win a Date with Tad Hamilton* the female star, Rosalee, has been loved her whole life by the person she thinks of as her best friend. He knows so much about her that he can describe her 'six different smiles'.

> ***'One when something flat out makes her laugh. One when she's laughing out of politeness. There's one when she makes plans. One when she makes fun of herself. One when she's uncomfortable. And one when ... one when she's talking about her friends.'***

Guess what's coming next? What are your six smiles when you deal with your customers? One when they first walk through the door. One when you're on the telephone. One when you're explaining your product. One when you're listening to a story they are telling. One when you're thanking them for their business. And one when what you're asking for is for use in referrals.

Stop! Wait a minute. I've seen people who smile all the time and look so fake that you don't trust them. *That's why a smile has to come from the heart and be sincere.*

So, what about the smile you use just because you are genuinely happy? When somebody smiles sincerely, you can see it in their eyes. The best way to practise smiling sincerely is to find some sincere things to smile about. Look at things that make you happy and when you feel a smile on the inside, show it on the outside too. If you're near a mirror take a moment to have a look at what you do when you smile sincerely. By having an awareness of how and when you smile, and by practising finding things to smile about on the outside, you'll find it easier and easier to smile sincerely, every day.

The **shortest distance** between two people is a **smile**.

17

One chance to make a first impression

The first thing that many women notice about men is their shoes. For the sake of a couple of minutes you would think more blokes would spend some time giving their footwear a freshen up. I know women who have walked out of restaurants, refused to buy clothes and turned away a rep because they had dirty shoes.

The first thing that most men notice about a woman is her smile (or lack of one). Men generally love a woman who smiles. Not in a sexual way but in a connective way. Most men may never look at a woman's shoes, but they'll always look for a smile.

This doesn't mean that men don't need to smile and women can have dirty shoes. The smile is one of the most important actions you can take, and it's simple.

After shoes and smiles there is a whole range of other impressions we give often before we've opened our mouths.

Tim Price runs Executel, a successful telecoms company. He loves to meet with sales people (don't all call him at once) and finds it fascinating how poorly most people make a first impression. *Here are Tim Price's top five observations of what not to do:*

1 Wearing what you wear to go out on a Saturday night to work. His basic premise is that you shouldn't try to be trendy for work. Tim points out that you can spend a small fortune on a shirt for a night out that looks great on a dance floor, but the ideal shirt for work should be completely indistinctive. If you don't give people anything to object to then they can't.

2 Colourful socks. I laughed until my sides hurt when Tim recounted a story of a 'big shot' sales executive who came to see him about a massive deal and, after a minute of conversation, crossed his legs and revealed his Homer Simpson musical socks. Classic.

3 Hair gel or over-styling. Again it goes back to the Saturday night vs work look (it's often young men who are the biggest culprits for this one).

4 Bad posture. Slouching, leaning to one side, resting on elbows all put Tim (and probably thousands of others) right off.

5 Bad breath or body odour. Seems very obvious but a packet of mints, a good scrub and some deodorant can go a long way.

I'll add three more which can turn a first impression into a lasting scar for me:

1. A weak (wet fish) handshake. A good firm handshake is essential. They should teach this at school.
2. Being late.
3. Being put on hold without being asked.

Actions

What do *you* think people should or shouldn't do when making a first impression? Write a list of 10 shoulds and shouldn'ts, then do them or don't.

18

I honestly don't care about your problems

A new restaurant opened in our local town. There was lots of publicity about the fantastic décor and background to the business. It was going to be the best place to eat and it would bring customers from the city. Fine dining like this on our doorstep was going to be wonderful.

Understanding that sometimes great restaurants take a couple of weeks to get into the swing of things, we avoided the opening night and the month that followed. So when the night of our booking arrived we were really looking forward to going out.

Other than the décor, which was really lovely, the night was a disaster. Warm (very expensive) white wine. Poorly cooked (very expensive) food. And staff who would have benefited from more thorough training. At the end of the night we noticed some friends of ours at a nearby table and joined them for a nightcap. Their experience had been similar to ours. Then, as luck would have it, one of the owners came to our table and asked how our night had been. This was one of those terrible choice moments: do you make someone feel better and say 'fine' or do you simply tell the truth?

We told her the truth. The lady listened, she nodded, said she understood and thanked us for the feedback. She said she hoped we would give them a second chance and return to the restaurant.

It was nice to have been listened to.

On the way out we met the other owner of the restaurant. Once again we were asked how our evening had been. This time we suggested he talk to his colleague as we'd just been through the whole story with her. But no, he insisted that we told him right there and then, every detail. He listened, shook his head and then went on to tell us a whole bunch of problems he was having. He griped about the supply of the wine, he blamed his staff (some who were just teenagers) and then he hit us with a classic. He explained that he was 'two chefs down' and one had walked out on him on Valentine's night.

At that moment I didn't care about why he had problems. I was **more**

interested in what he **planned** to **do** about them.

Here was a chap with a perfect opportunity to apologise, offer us a gesture of goodwill and ensure we returned to his restaurant to give him the opportunity to 'wow' us the next time. He could have made a new 'best friend'. He could have put a big deposit in our emotional bank accounts, but instead he took a withdrawal and with several hundred restaurants within easy driving distance we made our decisions and haven't been back.

How could he have improved our customer experience? Here's a simple list of do's and don'ts.

What *not* to say and do

Our system is down.

I tried but ...

We're having a staffing problem.

We've been very busy.

We can't do that because ...

Our policy is ...

The rules are ...

What to say and do

Thank you for your feedback.

We learn from what our customers tell us.

From what you have said we could be a lot better.

I'd like to make it up to you.

I'll make sure the right people get that feedback and we'll all learn from it.

Is there anything I can do for you right now to win back some of your trust?

19

Empowering staff

I was asked to speak at a conference for the BBC at London Zoo. The event went well and we had an opportunity to sell books afterwards. My first book *How to be Brilliant* had just been published and I was delighted when my team informed me that we had sold out (always good news for a first-time author).

I said my goodbyes and left to attend another event. Here's where a potential challenge occurred. Chris, the director of the department who had booked me, came over to the book stand, apologised for his delay and asked for a copy of my book. My team (by now one person) had to tell him that we'd sold out but we'd post one to him the next day. 'No problem,' he said, then turned to a colleague and said, 'Shame, that. I was going to read it on the plane in the morning.'

Here's where a bit of preparation and luck combine to create a brilliant five star service experience. My colleague, Nicola, was using 'heads up!' and heard his second comment, and then took action. Ten minutes later she was in a cab looking for a book store. Once she found a store she ran in and bought a copy of H*ow to be Brilliant*. Next she jumped into another cab and was on her way to White City (BBC HQ). When she arrived she persuaded security and the staff at Chris's office to allow her to place the newly purchased copy of the book on his desk with a handwritten note hoping he would enjoy *How to be Brilliant* and have a safe flight.

A couple of hours later Chris arrived to pick up his papers ready for the following day. Wow! was his first expression. He couldn't understand how a) we knew he was flying, b) we knew he wanted a book for the flight and c) a company based in Northumberland had managed to get a copy of *How to be Brilliant* to him.

And guess who he told about this? That's right ... everyone!

So how did Nicola know she could spend the time, money and energy doing this and why did she bother? Easy, since the early days of Michael Heppell Ltd we have empowered our team with details of: how much money they can spend, how much time they can take and that whatever they decide to do must fit with our company values, to make sure we create amazing five star experiences for our customers. You *never* have to ask permission to do this, you just go for it and, even if you make a spectacular mistake, that's perfectly OK. We present it as a triad to make it easy to recall.

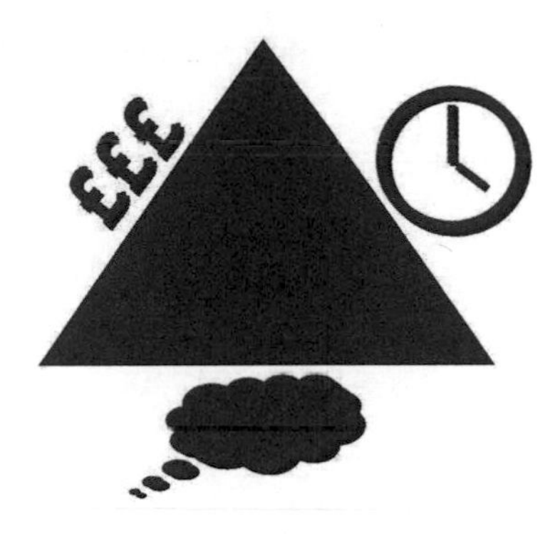

At Michael Heppell Ltd staff can spend one hour and up to £50 to make a customer's experience memorable and, so long as it fits with our values, it's perfectly acceptable.

Taking Nicola's example the formula worked perfectly:

- Total cost – £43.99 (taxis: £34, book: £9.99)
- Time – 50 minutes
- Values – Michael Heppell Value No.5 'Go the second mile and surprise'

Easy.

If you are the boss, ensure your staff know: **how much**, **how long** and the **spirit (values)** of what they are allowed to do.

If you are not the boss – ask yours to fill in the blanks.

How much money can your staff (or you) spend to make a customer's experience better? How much time can you devote? How will you know what to do and why?

Actions

Five things to do to empower five star service in your staff. If you don't have staff, ask your boss whether you can do these.

1. Set a budget – both in cash and time terms. It's amazing that people often don't take the chance to create an opportunity or fix a problem because they don't feel authorised to spend any of the organisation's money.
2. Share best practice. Create opportunities to share what others have done via awards/a notice board/a newsletter/team meetings, etc.
3. Look back at some 'live' examples of where people felt they weren't empowered and explore what would or could have happened if they had been.
4. Create a 'no fail' environment where, if someone should take an initiative and get it wrong, they haven't failed but instead they have had a big slice of learning.
5. Do it. Find opportunities to do it. Create opportunities to do it. Because the secret is not in the knowing, it's always in the doing.

20

Prepare for and relish competition

It wasn't until I got a new toothbrush and compared it with the one I'd been using for the last few weeks that I realised how worn out my old toothbrush had been.

And that got me **thinking**.

Later that day I bought two new tyres for my car and it was only when I saw the ones that had been removed next to the new ones that I realised how worn down they were.

And then that got me **thinking**.

That night I attempted to find my electric screwdriver in the garage and ended up throwing out and giving away four bags of things I don't use any more. I had empty shelf space and the garage felt clean. It wasn't until I cleared the shelf that I realised how much junk I'd been storing.

And that got me **thinking** and **thinking**.

We have never lived at a greater time for opportunity and prosperity. Take a look around. However, with this comes greater competition than we have ever before experienced. A shrinking world, superior technology, faster pace and higher expectations.

Competition comes in many forms. Competition from similar businesses. Competition for your resources. Competition for your time. (Are you a parent? You'll understand!) Competition in your relationships. Most people find a way to adapt, or get by, and do not realise there is a better way (or a firmer toothbrush!).

A friend of mine, a fine mechanic, opened a garage 10 years ago. He worked hard and built a good reputation. He had a steady stream of customers and did quite well. About nine months ago the unit next door to his became available and the landlord asked him whether he wanted it. He decided he didn't, said no and kept on doing his thing.

About seven months ago he called me and with a quiver in his voice told me that a new garage was moving in next door. It was opened by a younger guy, in his early twenties. The paint was fresh, the overalls were clean, the tools shone and, although he probably had one heck of an overdraft at the bank to pay for it all, it looked good, very good.

And looking that good, it made my friend's unit look bad – very bad. I didn't know how good this 'kid', as my mate called him, was going to be with cars, but he did have a whole bunch of technology in there to help him. I didn't know what his prices were going to be like in comparison with my friend's, though I did know he had some great opening offers advertised in the local paper.

But I did notice the change in how my friend operated his garage. I noticed the clear-out of all the junk and the new paint which quickly followed. I noticed how attentive he became to customers and I noticed how he fought for every piece of business that came his way.

And here's the best bit. Over a beer recently he told me how his new neighbour and competitor had said to him, 'If I'd known you were going to do all this I wouldn't have opened here, you just looked like an easy target.'

So with so many **opportunities**, yet so much **competition**, what's the **solution**?

There are lots, each complements the other and will make a significant difference *now*.

Actions

Six steps to help you prepare for and relish competition:

1. Become multi-skilled. By learning multiple skills you will always be in the right place at the right time ready to exploit every given situation. By developing a compulsion to learn you will also fine-tune your current skill set.
2. Make brilliance your benchmark. Brilliance doesn't happen by accident. When you set a new standard it automatically becomes a must to erase the old and replace with new.
3. As Lord Baden-Powell said, be prepared! Outsmart the competition with your amazing insight and immaculate preparation.
4. Aspire to greatness. You become what you think about the most. What are you thinking about? Focus on your success – feel it – create an emotive link.
5. Take a risk. If you're getting bull's-eye every time, you are probably standing too close to the target. Keep your risk muscle in shape by trying new things.
6. Get rid of excuses. When the Spanish explorer Cortez landed at Vera Cruz, his first instruction was for the crews to burn their ships. He then announced, 'You can fight or die.' This removed the third alternative of quitting. Remove your excuses.

So run with it. Get some new ideas, go on a course, listen to an audio learning CD, read a book – because you don't know what you don't know.

21

It's not what you say

'It's serious. Your car needs a new differential. We are talking thousands, not hundreds, and we can't do the work for a few days.'

It wasn't what I wanted to hear ... The service receptionist in the garage was just doing her job, but she could have learned so much from the technician (remember when they were called mechanics?) who had diagnosed the problem a few moments earlier. Here's what Paul Smith at Mercedes in Newcastle said: 'Mr Heppell, I've got some good news. We've caught the problem before it had a serious impact on your car. It's a new "diff" that's needed. It's not a big job to fit it, so we will see if we can get you booked in as soon as possible.'

'How much will it cost?'

'To be honest, it's not a cheap part but you have a really nice car and you look like a guy who wants things done right.'

So the technician told me the same message as the receptionist and added a few words that made me feel happy that my family and I were not in imminent danger. I felt reassured that it was the right thing to do and my car would soon be back on the road.

Fifteen minutes later the harbinger of doom gave me the same message and did all she could to make me feel low!

Why? My guess is that the receptionist has no idea of the impact she has on people when communicating in that way. She's 'just doing her job'. Or it could be that she likes the idea of a little bit of drama. Some people thrive on drama and negativity.

We've all heard the expression 'It ain't what you do it's the way that you do it.' This example is the same but this time it's 'It ain't what you say it's the way that you say it.'

I had a friend who was made redundant. When I asked how she felt, she said, 'I'm OK, but I'm worried for my boss, he couldn't have been nicer. It was really hard for him and he was so sympathetic. I'm lucky to have had him as my boss.'

I'm thinking, '**Wow**, what did he **say**?'

It turns out that the company had to make 20 people redundant to survive. Two managers each had to tell ten of their staff the news. My friend was let down by the nice guy. The other one had no idea how to give people bad news and ended up with four out of his ten considering unfair dismissal, bad mouthing him and the company, and trying to persuade the others to make a claim as well. Not one of the people who worked for my friend's boss wanted to take any further action.

What are you saying (and more importantly, how are you saying it) to your customers, colleagues, friends, family members? *Take time to focus on telling people news in a way that you would want to hear it.* It's not about lying, exaggeration or distortion. It's about giving people information in a positive way, in a way that you would find acceptable. In fact let's forget 'acceptable'; what about communicating in a brilliant way?

You will **feel awesome**. The people around you will **feel great** and, who knows, you may just get what you are looking for a **little bit faster**.

Wee Wow

Open your hands outwards when telling people news that could be better. It's good body language, which symbolises genuine regret.

22 It's your best friend – the awkward customer

You're doing a great job. You've done all you can and everything is sweet. Then suddenly it happens, the awkward customer appears. You know the one: they take hours of your time. They ask what seem to be the dumbest questions and change their minds constantly. No matter what you do, you can't seem to please them.

In *Fawlty Towers*, the brash owner Basil Fawlty dealt with customers, especially the awkward ones, in a unique way. He would scream and shout, and sometimes even physically remove them. He would treat his staff in the same way, and how we laughed. The temptation to 'do a Fawlty' is huge when Mr or Mrs Awkward is once again trying to make your life a misery, but you can't treat your customers that way. Shame ... but you can't.

Often the awkward customer has had something happen to them before you ever have the pleasure of dealing with them. It's rarely the fault of the person who is taking the wrath from this particular individual, however, they still get it double-barrelled, full on and with little consideration. This type of customer makes huge withdrawals from the emotional bank account (see page 47) of the unfortunate person who has to deal with them.

So what can you do?

Number one – you've got to listen. People love to be listened to, and awkward people really love it. Often they haven't been heard, so a good dose of listening can sometimes be all the awkward customer requires to get them 'on side'. Whatever you do, don't listen to that ridiculous advice about maintaining eye contact with the awkward customer. When you continuously maintain eye contact with somebody it can be quite disturbing – for both parties. The best way to show you are actually listening is to nod, repeat key pieces of information (at the right time) to show you've understood, and put your head slightly on one side.

They all help, but the **very best way** to show you are **actively listening** is to **take notes**.

It's amazing that when you take notes, people understand that you are taking notice. In most cases, that's what the awkward customer really needs to believe, that somebody is taking notice.

Number two – empathise.

This of course can be very difficult. Try using language like, 'that's awful', 'I'm really sorry about that' and then suggest a way to help. Often poor customer service operatives end up asking, 'What would you like me to do about it?' Don't do this as it is antagonistic and can often make the situation worse. *As the expert, you should know exactly what you can do about it.* Offer a couple of suggestions before you give the customer an opportunity to tell you what they would like. For example, 'Would you like a replacement product?' Followed by 'Or is there something else I can do for you right now?'

But what if they just need to 'vent' for a while? Let them. Some of us have a strong human need for significance. When these people come along, let them have their moments of significance by allowing them to have a huge download. Remember, what you don't have to do is take on board their negativity. In fact, you can turn it into a real positive moment – if you know how.

You mustn't feel you have to take the blame for somebody else's problems, particularly when you feel you've done everything you can within your power to make sure that their customer experience is five star. Remain calm and relaxed and in control. Stay creative and solution orientated to ensure you do the right thing during those stressful situations.

Bonus Bit

The next time you have a really awkward customer who just wants to give you a hard time; even when you know you haven't done anything wrong and they just keep on going; even when you listen correctly, say 'that's terrible' and offer them two solutions and the chance to come up with their own solution and if they still keep going – then self-preservation is required.

Picture them naked. See them getting smaller and smaller, and in your mind give them a strange, high-pitched, squeaky voice. This last-resort technique can be used to protect you from other people's negativity and reduce stress too.

Wee Wow

Three immediate actions to communicate bad news brilliantly:

1 Say sorry and mean it.
2 Offer a positive solution.
3 Follow up to make sure your customer is happy.

23

Building a customer service brand

Brand, brand, brand. It's all about brands in the business news today. Who has the most recognised brand, whose is worth the most? How much money was spent on a new brand launch? When it comes to five star service there are no prizes for guessing how important the brand is – it's everything. Because at the end of the day, the brand is you. It's what you do on a regular basis that makes the difference.

At seminars I often ask, 'Who would like £100,000 to be invested in them?' All the hands shoot up. Then I show a slide, which says, 'You Ltd' and I ask if they were a company and I had £100,000 to invest in them, would they deserve the money? Again lots of hands shoot up. We all like to think we *deserve* the money. When I ask why they deserve the money, we have the key moment. Most people have no idea why and the hands start to drop. The ones who remain are asked to give a 15-second commercial as to why they should get the cash.

Their replies are varied. From the people who glibly offer to double my money (I still can't get them to put that in writing), to those who give a pitch as though they're a registered charity and before they've had a chance to whip out their violins they've usually run out of time. After four or five 'infomercials', I ask the group who they would invest in. In most cases I get a response which exemplifies what being 'brand you' is all about. It's the most likeable person who gets all the votes. Not the ones who 'deserve it', not the ones who insist they are a 'great investment', but the ones you would most like to be friends with.

And that's what customer service branding is all about – making **your service so good** that you end up being with **friends**, not just **customers**.

Kevin Roberts (the worldwide CEO of Saatchi & Saatchi) talks about 'lovemarks'. He describes them as 'the future beyond brands'. A lovemark takes loyalty to a whole new level where not only are you loyal to the brand, but you actively promote it. You persuade others to switch, you feel like you are part of the family and you fight their corner. Most lovemarks are created by organisations that care about people first. Could you describe your organisation as one worthy of a lovemark?

BIG QUESTION

If the brand is all about **you**, what are you doing to grow '**brand you**'?

Actions

Here are five thoughts to grow 'brand you' and make your customers fall in love with yours truly:

1. Remember, you are the brand. It's just the company logo that goes on ads and stationery.
2. Brand needs constant rejuvenation. Never allow yourself a day off from being the very best you can be.
3. Brands are built on rock-solid values. Do you have values written down that you stand by every day?
4. Brands are borderless. Does what you do extend outside your immediate circle of influence? Could 'brand you' go international and still get top results?
5. Question why you are loyal to some brands and repelled by others. Take the parts that attract you and do them more, take the parts that repel you and do them less.

24

Be individual, encourage individuals

I am not an airline snob as I just want to get from A to B in the quickest and most convenient way, so when low-cost no-frills airline easyJet started to fly out of my local airport I was thrilled. As my dad said, 'In business you never make money faster than when you're saving it' (it sounds perfect if you read that with a deep 'dad-like' voice in your head).

On an easyJet flight from London to Newcastle I had one of the best experiences I have ever had. Why? Because of Simon. Simon is the type of air crew you would love to headhunt for your organisation – even if the only purpose was pure entertainment value. He turned a dull 45-minute trip into a memorable experience. How did he do it?

From the moment we stepped on to the plane Simon had a different way of welcoming almost every person. He knelt down to greet the kids, he bowed respectfully yet humorously at two Japanese businessmen and, as I boarded, he said, 'I'd love to help you with your bag, but you look much stronger than me.'

But Simon came into his own once he got hold of the microphone. In a wonderful perky voice he announced, 'Welcome to this easyJet flight to Geordieland.' Most people cheered, although there were one or two groans. Quick as a flash he said, 'Don't worry, we will get the Mackems home too' (Mackems are people from Sunderland, Newcastle's big rival city 15 miles away).

He then went on to say, 'Please put all items in the overhead lockers. The only things that should be on the floor when we take off are your feet and the carpet.' This was closely followed by my favourite, 'Please turn off any electrical items, especially mobile phones, as they could interfere with the aircraft's navigation systems. The last thing we want to do tonight is end up in Carlisle.'

Once we were off the ground the cabin crew did a quick drinks trolley run where Simon offered everyone a 'cocktail of their choice'. The chancer sitting next to me asked whether he could have a pina colada. 'No problem at all, sir. But it may taste like lager and tonic.' And once again everyone howled with laughter.

When we landed at Newcastle Simon was ready for his grand finale and did the whole landing welcome in the style of Lily Savage. I had to write it

down as it was so good. Get a hard Liverpudlian accent in your head and read on, it went a bit like this: 'Hello, ladies and gentlemen, and welcome to Newcastle. Please don't even think about getting out of your seats until the plane has come to a complete standstill and that light has gone off. I'm watchin'. Please be careful taking any items out of the overhead lockers because after a landing like that they're bound to have moved!

'It's cold and damp in Newcastle tonight – for a change – so be careful as you go down the steps as they could be a bit slippy, love. Please make sure you take all items with you but don't worry if you forget anything because me and the girls are going to a car booty on Sunday and we're looking for a bit of extra tat. We realise that you have a choice of airlines. I can't believe you chose easyJet but as you did I have to say on behalf of all the crew, we appreciate your custom.' Then it happened. For the first time ever on a plane that hadn't been saved from certain disaster by a brave pilot, everybody burst into applause.

Simon created magic moments for everyone on the plane and judged the humour level brilliantly.

You don't have to be just like Simon.

But let's take a quick look at what he did and what you can do too:

1. Know your customers – he had a different way to welcome most people on to the plane.
2. Know your product – he'd obviously made his announcements dozens of times and knew what he was doing. This gave him the confidence to do his extra material.
3. Local (or specialist) knowledge – he knew the local terminology for the people who lived in the major cities, he knew about the reputation Newcastle has for being cold, he knew what he had on his drinks trolley.
4. Get everyone involved. It is great to create a five star moment for an individual but to do it for over 100 people – that's special.
5. Use humour. Don't you just love to laugh too?

The difference between one, some, all and many

In my book *How to be Brilliant* I wrote about Harry Nicolaides who worked as a concierge, and the truly brilliant influence he had on me and the guests of the Rydges Hotel in Melbourne. It is with deepest regret that I must inform you that Harry has moved on to live in Phuket, Thailand.

The last time I visited Melbourne I arrived in Rydges with an air of anticipation, wondering whether the experience would be the same without him. Within 30 minutes of arrival it hit me. Rydges was now another average hotel. Why? Brilliance was the culture of one or two individuals and not of the organisation as a whole. I was devastated! The first time I visited I was on my own, this time I had my family with me and I hadn't shut up about how amazing this hotel had been. Now that Harry had moved on, sadly the magic had left with him.

So how do you **create** a **five star culture** rather than relying on one or two individuals who make your organisation shoddy or **shine**?

Get the values right. Service values are critical to the success of any organisation. Make sure they are written down, widely distributed and most of all used. Without the right value system, how will you know whether you are creating a five star culture?

Almost all organisations have a mission statement or vision – it's usually written by a bunch of executives during an 'off-site meeting' in a large country house. Values are different. They need everyone to have an input because when you really live them (especially service values), decisions are easier, people are clear on their wider objectives and customers really feel them. *Is five star service a company value for you?*

Have you ever thought about arranging an internal training session to work on this? It can be as easy as getting a team together and asking everyone this key question: 'What was your best customer service experience and how much did it cost?' You'll soon notice that the very best customer service experiences have little or no financial cost. The next step is to ask what you have learned from people's anecdotes and which bits you can apply in your organisation.

Encourage some team members to take up the challenge (or do it yourself) and create posters, stickers and devices to promote the level of customer service you expect. *Celebrate success.* Read 'Ring the bell' on page 61 (about celebrating successes). This gesture makes everyone in an area look up. It encourages people to share success and, if you haven't rung the bell for a while, you'll do whatever it takes to make it ring!

You may be a service genius, your organisation may have two or three service superstars and that's great. But if you can't create a culture where five star service is a value and a standard, you and the other stars are going to become frustrated very quickly. So it is well worth the effort to get others involved.

Is it worth it? Well, we're living in times where customers demand more from less and they don't even tell you about it. You must be at the leading edge when it comes to creating outstanding five star customer service. Make it a must!

So how do you write a set of service values you can live by?

First of all, writing any set of values is as easy or as difficult as you'll want to make it. I've worked with companies that have done it in an afternoon and been very successful and I've worked with others that have taken months and barely got off the starting block.

I'm going to assume that you will be the champion of your organisation's company service values. Here's a simple system you can use.

Get everyone (or as close as you can) together. This is probably the single hardest task. You'll have to find a way to get maximum buy-in to the process and getting together the people who are going to be delivering the message is vital. If you can't then find a powerful way to communicate what you are doing to the whole organisation. This needs to be more than a memo.

Introduce why you think having customer service values is important. Give some anecdotal evidence and share the benefits of values: freedom, compliments, ease of decision making, etc.

Now ask your team two questions: what is **important to them** and what do they think is **important to their customers**.

It's often a good idea to get small groups discussing these ideas rather than trying to take lots of feedback from a large group (if there are fewer than 10 people in your organisation, that's a perfect number to work with).

Look for themes

Look for themes running through the information given but be aware of not leading the group down a route they don't want to go. Identify words that can be used as values. Say you have three statements saying, 'Wow our customers', 'Make it feel new' and 'Be different'. Ask the group whether they think the word 'surprise' could sum that up. If they do, great; if not, ask why – and listen.

Key words

The next stage is to work towards getting a few key words that sum up your customer service values. Here's where it can get interesting because people often want to put the cart before the horse and know an exact description for a value before it goes on the list. That's when it takes weeks instead of hours, so tell the participants you'll come to that but right now you are looking for some key words to work on.

Prioritise

The next stage is to take what could be a long list and make it into a shorter one. The simple way is to look for similar words and chunk them. Then prioritise them and lose the bottom ones if it feels like you have too many (rule of thumb – if you can't remember the list you have too many).

Write descriptions

Then you write the descriptions. Again this could be done in small groups if you are a big team or as a whole. Descriptions should be short and concise and give a brief empowering description.

Spread the word

When you've done all that make sure everyone knows them and knows how to use them and encourage everyone to sign up to them. Next print them out and stick them on the walls. Make small laminated cards with them on. Tell people to share moments when they use them – challenge the people who don't. Post them on your website (you can see mine at www.michaelheppell.com). Make them a part of your culture. This takes a little bit of time and may never be perfect but the effort pays back substantial rewards.

If you're stuck while brainstorming, here are a few ideas to get you started:

Brilliant	Accurate	Polite
Punctual	Magic	Creative
Fun	Thoughtful	Unique
Involved	Excellence	Referable
Genuine	Friendly	Fun
Best value	Motivated	Surprise
Second mile	Clean	Famous
Systematic	Consistent	Funky
Award winning		

26

Recruiting service professionals

Hope is not a strategy when it comes to recruiting staff with a great service ethic. In fact neither are most of the diagnostic tools that may assess many things, but rarely predict how great staff will be with customers.

I remember once watching a programme in America called *The Rebel Billionaire*. It was a show a bit like *The Apprentice* but it featured Richard Branson in his quest to find someone to take over running part of Virgin. In the first episode the 16 young rebels flew in from the States, bright-eyed and ready to take on any challenge. They arrived in groups of four and were picked up in a London black cab. A film crew was on hand to capture them climbing into the cab then, unbeknown to them, a secret camera recorded their journey to Branson's house.

So far so good – you couldn't have messed it up already? Could you? At this point you haven't even arrived at the interview.

The first thing they failed to notice was the elderly taxi driver and how he was struggling with their cases. Did they stop and offer to help? No, in fact a couple of the blokes let the girls struggle too.

Next was the conversation they had in the cab. Granted, the old taxi driver was egging them on by asking what they thought of Sir Richard but it was amazing how freely they shared their views, positive and negative, with a complete stranger.

Once they had all arrived, they met in the lounge and waited expectantly for their introduction to Sir Richard. Silence ensued as they did their best to look cool and professional. After all you get only one chance to make a first impression.

It was amazing to watch their mouths open when in walked not Sir Richard but the old taxi driver, who proceeded to stand up straight, change his voice and tear off his bulbous nose. Yes, you guessed it: the taxi driver was in fact a brilliantly made-up Sir Richard Branson.

He went on to show them some video clips of their behaviour since arriving at the airport, which left several of them red-faced. For one participant, who had made a particular idiot of himself, it was time to use the return portion of his ticket.

When it comes to recruiting service professionals you must **recruit on values** and **attitude first**.

Skills can be taught quickly, attitude takes longer and personal values are developed over years. I think you know the sort of attitude (or attributes) and values I mean: energetic, polite, caring, clean, timely, etc.

Here's a simple grid to help you decide, following your first encounter, where a person may fit with your five star service culture. It's not the most scientific approach but by scoring 1–10 and plotting where they land you may find you a gem.

Skills	LOW Attitude & values	HIGH Attitude & values
HIGH	Can this person change?	HIRE THEM NOW!
LOW	Thanks but . . . no thanks	Give them a chance to learn the skills

The actual interview process (I think) is simple. Mainly because in an interview you have almost certainly made up your mind in the first few minutes and you then spend the rest of your time justifying your decision!

HR professionals may disagree. Here's what Liz McGivern, HR director of the multi-award-winning Red Carnation Hotel Group, says about what to look out for when interviewing service professionals.

First, I wouldn't say *don't* listen to your instincts as some people have good ones! I would say, don't *only* rely on them! A slightly more scientific approach could involve the following:

1 Ask yourself when you meet the candidate and during the interview, **what outlook or attitude does the candidate have**, particularly when you first meet and shake hands? (Is the candidate a natural smiler, friendly, who makes good eye contact with you? Or perhaps an extrovert and clearly interested in the job?)

2 These are all great pointers on how that person would be with your guests so take note and see if you can picture them working for you. Remember even if they are nervous (and you can allow for nerves to an extent for some of your entry-level positions) **you are seeing them at their very best**. You could be compromising if you settle for less than these personality traits.

3 Asking questions on past experience is useful though you must ensure you are looking for actual episodic memories in the candidate's answers and not what I call 'textbook' or hypothetical answers. For professional service positions I like to **ask questions about a time they have dealt with a complaining and possibly angry guest**. What had happened? What did he or she say to calm the guest down? How did the guest respond to them and what did he or she offer to do for the guest? What was the outcome? Someone who handed the problem over immediately to their supervisor may not be able to act on their own initiative. On the other hand you may hear a brilliant story of how your candidate thought of and did all sorts of things to ensure the guest went away happy.

4 Other questions I ask for service positions could include **an unusual request and how they dealt with it**; a time they had to make a last-minute change; **a time they had to break the policy or procedure (or at least bend it a bit!) to give better service**; his or her best sale; a time when

they turned a negative situation into a positive one. Answers that are based on actual situations candidates have been in are revealing and helpful to you and indicate how someone is likely to behave in the future.

5 Last, whenever possible, invite **someone in for a trial day, or morning, or hour**! It's a two-way street for potential employee and employer. Remember, again, that as the employer you are seeing the person at their best for the time they are with you.

So whether you're a professional recruiter or someone who just needs a part-time pair of hands, your intuition is the most important part of the process. Ignore it and you may just end up with second best.

27
Super scripts

Remember the great treatment I received from the receptionist who took my credit card at the Ritz Carlton in Singapore? If you don't, here's a reminder.

> 'Mr and Mrs Heppell, you are going to be staying with us for the next five nights. During your stay you may wish to order room service or perhaps have a drink from the bar. You may wish to purchase something from one of our boutiques or join us for a meal in one of our fine dining restaurants. Perhaps it would be convenient for you to allow me to take a swipe of your credit card so you don't have to worry about carrying money.'

I don't know how many times a day she says that, but I do know it's a fantastic script. She may vary the odd word or two but basically she says that because it works.

The pros and cons of scripts

Scripts make customer service easier when they are delivered **passionately** and **sincerely**.

You say the right thing, you say it effortlessly and in most cases you know what the results will be.

Scripts give continuity. Great to take a person from A to B. Scripts give confidence. 'I didn't know what to say so I just said nothing' was what one person told her boss recently.

And for those of you who think scripts take away passion, think about watching a Shakespeare play ... passionate?

But be careful. How many times have you heard a script delivered in a monotone from a person who, if they weren't standing in front of you, you would swear was reading from a screen? (In lots of call centres they do!) Just because we can't see someone on the phone doesn't mean we can't feel them (see 'Telephone service' on page 129).

Scripts can kill creativity. If you are using a script and you need to digress from it to say the right thing for your company and the customer, then *go for it.* Even the best script can be improved to get the most out of a situation.

The following is an example of where a script works brilliantly. Just imagine you have booked a holiday and the travel agent says to you:

> 'In the next couple of weeks you'll be receiving your tickets followed by final confirmation of your flight times about two weeks before you fly. This is my card and there's my number if you need anything.'

That would be nice. Or what if they said:

> 'Well that's you all set. In the next couple of weeks you'll be receiving your tickets, so look out for the postman. Then you'll receive final confirmation of your flight times about two weeks before you fly, that's when things get really exciting. If there's anything at all that I can do to help in the meantime, remember my name is Chris, this is my card and there's my number. You guys are going to have a brilliant time.'

That's five star!

Actions

Three keys to a great script:

1. Write it down. Think about the words you like to use and the way they will flow.
2. Read it out loud five times. Make small adjustments each time until you like it. It will still feel false the first few times you use it but once it's committed to memory it will flow easily.
3. Be prepared to deviate slightly to say the right thing at the right time but generally stick to the basic script.

28

Voicemail, answering machines and automated call queue systems

**ow many times** have you called somebody and heard this kind of message:

> 'I'm either on the phone or away from my desk right now. Please leave a message and I'll get back to you as soon as I can. Please speak clearly after the tone.'

Are you really going to set the world alight with that one? Why not create a voicemail that promotes what you do in a five star way and at the same time does the important job of encouraging people to leave a message?

Imagine if you were a plumber and your voicemail said, 'I'm sorry I can't answer the phone personally right now as I'm out on a job installing some quality pipework for a valued customer. Leave me a message after the tone and I could do the same for you.'

What if you were a photographer who specialised in portraits and your voicemail said, 'Right now, I'm helping somebody capture a family image so they can treasure this moment for a lifetime. Please leave your message after the flash.'

What if you worked in an accounts department and the message said, 'I'm sorry I can't take your call right now as I'm taking a break from my desk to make sure that when I come back I'm fresh as a daisy and able to make the numbers match, the payments accurate and get the invoices out on time.'

I would love to call a company that has a queue system that starts off with, 'You've probably noticed it's one of those system things where you have to press a button just to get to the next menu. Well, we do this because when it works it's much more efficient and honestly it does benefit you. So here we go, time to make your first choice, and press one ...'

Recently I've noticed that some people are leaving unique messages, especially on mobiles. That's a great idea because it makes that person seem very real. However, they then say something like, 'It's Monday the 21st, I'm in meetings most of today so please leave a message and I'll get back to you.'

Wouldn't it be brilliant if instead it said,

> 'It's Monday the 21st and today I have the privilege of meeting with some customers in the morning and after a quick lunch I'm going to a product launch to see the latest software designed to enhance our business. So don't hang up. Instead leave me a message and I'll get back to you either later on today or first thing tomorrow morning.'

I know it's not suitable for everybody reading this book to have a message like that, but I would bet that over half of those reading could.

Be careful though. Don't make your message so long and entertaining that by the time your personal message ends the caller has hung up. Be aware how often you get the same people leaving you a message – if you get the same people all the time they may just need a couple of words. I have a friend whose message says, 'Dan. Speak at beep.' Dan is a man of few words but cheap mobile bills.

Go on, I dare you to rerecord your voicemail message right now, have some fun and take a risk.

Bonus Bit

Here are 20 great words to use in a message (not all of them!):

Promise	Missed	Quickly
Grateful	Out there	The best
You	Short	Personally
Changing	Making	Hi
Please	Clearly	Brilliant
Thank you	Magic	Appreciate
Earliest	Rapid	

Wee Wow

'Out of office' replies on emails are often written with about 10 seconds' thought and 30 seconds' typing time (I know it's because you are usually about to go on holiday – yippee!) but they should always have a '*what to do* and *who to contact'* if it's urgent part. You don't want your customers wasting their time or your organisation's time by having to call in or, worse still if they can't get you, could they call your competition?

29

Telephone service

'It's a brilliant day at Michael Heppell. This is Sam speaking, how may I help?

That's how my staff answer the phone every time you call us. Cheerful and upbeat. It immediately lifts people. It sets the tone for the rest of the conversation. It gets us noticed. I love being in our office and hearing people answering the phone. I love ringing the office and hearing people answer the phone.

Then when we call other organisations ... you know what it's like. Often it's the company name blurted out, followed by a couldn't care less 'How may I help?'

'Only One Star, how may I help?'

But usually it's just the company name.

'Only One Star.'

Dull, dull, double dull and not very five star. So step one is to work out how you are going to answer the phone. It doesn't have to be 'It's a brilliant day ...' but why not choose a much better welcome than you are using right now?

Here are a few thoughts for our fake company called Only One Star.

'Welcome to Only One Star, this is Heather, how may I help?'

'Hello, this is Heather at Only One Star, how may I help?

'Thank you for taking the time to call Only One Star, how can we help you today?'

If you called Only One Star, which way would you like the phone to be answered?

Now we've answered the phone, what's next?

Have you ever had a telephone conversation with someone and they just didn't sound interested in you? That's because they weren't. They may have said the right words in the right order but their body language and tone told a different story.

And you heard it.

Just because a person on the phone **can't be seen** it doesn't mean you can't **feel them**.

In fact, when talking to someone on the phone it's best to exaggerate your physiology (what you do with your body) and slightly exaggerate your tone. Here's how to do it:

1. Sit up or, even better, stand up.
2. Look slightly upwards.
3. Smile – smiles can be heard over the phone.
4. Stay focused – no matter what else is going on around you.
5. Create a visual image of the person you are talking to – make them look happy and nice.
6. Use the person's name.
7. At the end of the conversation remember to say a sincere thank you.

Do those seven simple steps on every call you make and you'll quickly see the difference in the quality of your conversations and how your customers respond to you.

But if you want to really excel, dive into 'Advanced telephone service' in the following chapter.

Test us out

If you want to call my office and hear my team answer with 'It's a brilliant day at Michael Heppell ...' etc then be my guest but please say hello to my team member who answers and tell them you are reading *5 Star Service*.

30

Advanced telephone service

We've got the basics of creating a five star telephone style, so what's next? Great five star telephone service practitioners have habits that others can only dream about. At first they seem a little complicated, but very quickly you'll get used to them.

Voice mirroring

You may have heard of 'matching and mirroring' as a way to build rapport. It's a great tool to use but most people limit it to face-to-face meetings. On the phone it works just as well.

There are three important actions to mirror on the phone:

1 Key words.

2 Personal phrases.

3 Visual, auditory or kinaesthetic cues.

Let's start with key words and personal phrases. As you are engaged in a conversation, make a note of any key words the person uses. You'll know them as they'll stand out of the conversation. When you talk to them your task is to replay at least three of the words or phrases.

Here's an example where a business traveller is looking for a room in a hotel.

> 'I'd like a nice big bed in a quiet part of the hotel, please. I'm not bothered about a view as I'll be working and having a bite to eat in my room. Do all of your rooms have plenty of sockets?'

Our traveller rang two hotels with the same request. Read the two responses and guess which one got the business.

> 'We have a queen-bedded room available in the west wing, our "Guestronomic" room service is available 24 hours and every room has full internet access and 240-volt power supplies.'

Or

> 'I've got a room here with a lovely big bed in the quietest part of our hotel. We'd be happy to serve you a bite to eat in your room and I can confirm all our rooms have plenty of sockets.'

As you can see, the first person gave all the correct information but it was right out of the brochure. Our traveller isn't bothered about a 'queen bed' or that the room service is called 'Guestronomic' or that the voltage is 240. The second person listened, noted and replayed several key words and personal phrases. Our traveller used the term 'lovely big bed' so got 'a lovely big bed'. And used the term 'bite to eat' so is going to get 'a bite to eat'. See how many more you can spot in those two simple sentences.

Not only is this very powerful, it's also great fun. Give it a test – you'll love it.

Here are some classic mismatches (both on and off the phone) – how we can get the subtle things wrong and what we could say:

Customer	**Wrong Answer ✗**	**Right Answer ✔**
I need it urgently	We'll send it quickly	We'll send it urgently
Can you fit me in?	We have a vacancy	We can fit you in
How much are the fees?	The cost is	The fees are
I'd like it in scarlet	We have it in red	We have it in scarlet
Does it contain dairy?	There's no milk in it	No, it doesn't contain dairy
Can I return this?	We can take it back	Yes, you can return it

Primary styles

The next area to look at is using and understanding people's primary styles. Most people fit under one of three primary styles:

Visual

Auditory

Kinaesthetic

When you're on the telephone to a visual person you will recognise that they tend to use visual 'cues'. For example, they will say things like 'I've seen your product and it looks very interesting, could you send me a brochure so I could look at it in more detail?'

When you're on the telephone to an auditory person you will recognise that they tend to use audio cues. For example, they will say things like, 'I've heard about your product and it sounds very interesting, could you tell me a bit more about it, I'd like to hear some more detail.'

When you're on the telephone to a kinaesthetic person you will recognise that they tend to use kinaesthetic cues. For example, 'I know about your product and it feels like something I could be interested in. Can you describe what it does and keep me in touch with future development plans?'

You will also notice that the three types of people tend to talk at different rates. Visual people tend to talk very quickly. You've heard the expression 'a picture is worth a thousand words', well visual people try to fit in all one thousand words as they picture what they would like to say.

Auditory people tend to be more paced and often have a wider vocabulary.

Kinaesthetic people tend to be very slow and often they will pause as they try to get in touch with their feelings. At this point auditory and visual people tend to jump in and try to finish sentences for them. Kinaesthetic people really hate this so if you are talking to one, let them finish.

You can have a lot of fun by attempting to understand primary styles early in a conversation, then communicating back to the other person in that particular style for the rest of the conversation.

So why should we take time to learn and apply these ideas?

Basically **people like people** who are like **themselves**.

In simple terms, if we know what these primary styles are and use them, at a subconscious level the people we are talking to think that we are more like them. And if we like people who are like ourselves, we create a much better customer experience.

Like learning any new skill, the more you do it, the easier it gets and quite soon you'll be able to mirror a whole range of people's tones, styles and nuances.

Checklist of words and phrases to look out for:

Visual	**Auditory**	**Kinaesthetic**
See	Sounds	Feel
Look	Listen	Hard
Picture	Hear	Get a handle
Vision	Deaf	Touch
Reveal	Rings a bell	Grasp
Imagine	Silence	Concrete
Show	Deafening	Stroke
Clear	Tune in	Finger
Image	Hush	Gut

At the end of the day most people aren't totally visual, auditory or kinaesthetic, they tend to be a mixture of VAK (visual, auditory and kinaesthetic), but many do have a primary style which they use to communicate with, and this is the one to listen out for, feel and focus in on.

31

Customer magic moments

Customers need to have **their magic moments** too.

Even if it's no big deal to you, it may be incredibly important to your customer. You can't decide how significant it is but you can decide how much attention you will give it.

I was waiting to meet a client in an overly expensive London hotel when an American lady came through the front door gushing over her new discovery. She grabbed the concierge and proceeded to tell him how a wrong turn became a fortuitous mistake. This is what I heard.

> 'Can you believe it; I took a wrong turn on the way back to the hotel and ended up in the middle of nowhere!'
>
> The concierge didn't do anything to hide the look on his face which screamed, it's unlikely you could end up in the middle of nowhere in Central London.
>
> The guest continued, 'But I'm so glad I did get lost because I discovered an amazing little park.' She pulled out her map and said, 'Look it's right here!'
>
> Now at this point the concierge could have said 50 different things but what he came out with astounded me. He looked at her map and said, 'I know, I walk past there every day.'
>
> Instantly the excited guest looked forlorn. He'd burst her bubble and taken her moment.

So what could he have said? How about, 'Let me see … wow, what a gem, and so close to the hotel! I walk to work that way, I'll have a closer look tomorrow. Thank you.'

Then can you imagine how the guest might have felt if the next day she had a note in her room thanking her for the recommendation and adding he had followed her instructions and taken a walk past the park that very morning.

I continued to watch this concierge for the next 15 minutes while he recommended routes, booked tables, gave advice on opening times and offered the low-down on the hot spots in town. All of which made him look good, polished his ego and made him feel fantastic. Shame he couldn't do the same for his customer.

Here are the 5 star ways to make your customers feel magic.

1 Remember, even if you've heard the story a hundred times, it's your customer's first time.
2 Actively listen. Nod, ask questions and make the moment real.
3 Follow up. Add something to make them feel even better the next time you see or talk to them.
4 Share with others what you have gleaned from your customer and credit the source.
5 Be in the moment. Easier said than done. This means you are 100 per cent focused on them. Not checking emails while on the phone. Not looking over someone's shoulder. Not planning what you are going to do next or replaying what's just happened. It means you are there for them and in their moment.

32

What's in a name?

Everything!

If you use people's names on a consistent basis and really care about that person's name, you could receive rewards beyond anything you could have imagined. It's pretty difficult to get sick of hearing your own name – in fact it's almost impossible.

The two replies to the question I ask below could be directed at anybody, but for the sake of the example the person I called was talking to me, 'Michael' or 'Mr Heppell'.

> 'Hello, it's Michael Heppell here. I'm calling to ask if you know when my shoes will be ready.'
>
> 'Just let me check. They are going to take about three more days, is that OK? Is there anything else I can help you with?'

Or

> 'Hello, Mr Heppell, thank you for calling. Do you mind if I call you Michael? I'll just check on your shoes right now. (Pause while checking.) OK, Michael, I see it's going to take about three more days, is that OK? Is there anything else I can help you with today?'

You could almost drop in a person's name every other sentence and it wouldn't be too much. Plus, when you use a name, you remember a name. And that's really cool when someone remembers your name and you weren't expecting it. (See 'Putting on the Ritz' on page 19.)

But what if you don't have to speak it, you just have to write it? I get my surname misspelt all the time. Heppel, Hepple, Heppal, even Hepull are some of the favourites. I love it when someone asks, 'How do you spell your surname, Mr Heppell?', then reads it back to me. I have a close friend whose surname is Franckeiss. If he had a penny for every time someone misspelt his name he'd be a millionaire.

As we are living in a multi-cultural society we are experiencing a wave of new names and spellings into our vocabulary. Most people don't mind if you ask for a pronunciation of their name or a spelling. If I hear a name for the first time I ask, 'Where does that name come from?' and 'What does it mean?' If a person has gone to the trouble of finding out what their name means then, trust me, they love to tell you.

As it happens, most people don't offer you their name. Make it your mission to find out. I make a conscious effort to find out the names of personal assistants and I work hard to be genuinely interested in them. People feel insincerity, so if you aren't genuinely interested don't try to fake it – you'll get found out.

Here's a typical call. 'Hi, Wendy, how are you today? Still as busy as ever?' Then as soon as I get a reply I use their name again. 'Wendy, I was wondering, is it possible to get an appointment with Sue at some point early next week? I know she's busy, Wendy, but if anybody can get me half an hour it's got to be you.'

I always make a conscious effort to use names when I'm looking for five star service for myself too. I know that people treat their friends better than they treat strangers, and we tend to call our friends by name. Request our bonus chapters (see page 236) and find out more about how you can become a five star customer.

I don't know about you, but I love going to those events where everyone is wearing a name badge. However, it still amazes me how many people don't actually read the name of the person they're talking to. You don't even have to have a good memory and you can glance, pick up someone's name and be using it within a few minutes. Easy.

Companies often have their directory online which lists the names of all the people you might want to talk to. So when you call you can ask for someone by name – it's much more powerful to ask for someone by name and get the right person than it is to ask for a title or department.

There is no **sweeter sound** for someone than the sound of **their own name** – so use it.

Actions

Two simple ideas for learning and remembering someone's name when you first meet or speak to them:

1 Use repetition and say their name at least three times in the first few minutes. By saying it three times you are 60 per cent more likely to remember it. Say it several times in your head too, that also helps.
2 Create a visual image that links their name to a prominent part of their appearance. The more bizarre the better for this as your brain loves colour, humour and the unusual. So, for example, if you meet someone called Peter Green and he's wearing a big coat, imagine a bright green 'pea' that creates a 'tear' in his coat. Bizarre, I know, but at the end of this book I bet you remember Peter Green before you remember any other names mentioned.

Wee Wow

Find out the meaning of the 10 most popular names and use this information in conversation. It's a nice idea to have them written down and available if you use the phone a lot.

33

Know your competition

Who's your competitor? When was the last time you asked yourself that question? Perhaps you currently invest in the latest business research to establish who your competition is and how well you stack up.

As an ambitious individual you may take a look at your colleagues and work out whose abilities match or outweigh yours. Or maybe you haven't really thought about it at all lately.

To be honest it doesn't really matter whether you know or not because the real answer is . . .

Everyone!

Yes, we live in a world where we truly are in competition with everyone else. And how do I know?

Because as a consumer I don't compare like with like any more. **I compare experience with experience.**

I compare the way one group of people meets my needs to another group; the way one organisation makes me feel and how another completely different organisation makes me feel.

It's not fair, but it's true. You do it too. Have you ever had an experience in a shop and wished it was more like 'X'? Or had an experience with your garage and wished it was more like 'Y'? Bet you have.

I was sitting on a Virgin train recently when the chap who makes the announcements for the on-train shop, 'Gary Lucky McLuckie', had the whole train salivating at his buffet announcements and laughing at his wit. In my carriage two people commented, 'I wish more people were like that.' I had to ask them, 'Don't you wish you were more like that?'

The next time I hear any announcement from any person in any environment I'll compare it with Gary Lucky McLuckie's. Not fair, but true.

This means you're in competition with everyone who:

- picks up the phone faster
- delivers an order more quickly
- exceeds expectations more regularly
- has a better tone on the phone
- lives values more passionately
- understands customer needs more clearly
- goes an extra five miles when you went an extra one.

How does it feel to be in competition with the best in the world at everything? Tough, eh!

But here's the good news. You now know that you are in this crazy competition – and the majority don't. It means that you have the magic of a 'first mover' advantage and unless your competition is reading this you can take massive action to make a difference now.

We tend to put the word 'service' as a big banner over such a lot these days it's not surprising we compare all these experiences in a big melting pot. The key is to know what to do with it.

Let's take it to the next level. If our customers no longer compare like with like (and why should they? You don't) then where might we end up?

Say you work in a shop selling clothes. In walks a customer who has just flown with an amazing airline, been picked up in a beautiful clean cab by the friendliest taxi driver, checked in to a great hotel where the staff have such ESP that guests' needs are met before they knew they had a need, and then they come to see you. What are they thinking?

That's right – their expectations have been raised. If you are anything other than five star you'll let them down. But you don't know that. You don't know about the plane, taxi or hotel. You don't know where they are but I bet you they'll compare your service (even if it's only at a subconscious level) with their recent experiences.

And here's where you can win. Most people who walk into your shop won't have had a brilliant flight and a fantastic cab ride followed by amazing service in a hotel. They'll be hassled to bits, have found it hard to park, run in during a lunch break and their recent experience of customer service will be poor. That's why when you give them five star service they'll compare you with the rubbish they've just experienced (because we do) and you'll shine.

Four big points

Understand this.

Most people compare experience with experience, not sector with sector or product with product.

Most of your competition are rubbish at customer service.

For you to be brilliant you have to give five star service *all the time*.

You *will* be compared to the others, but the good news is most of what you are being compared to is poor – so when you shine you'll be seen as amazing.

34

Speed it up!

What do we want? Five star service!

When do we want it? Now!

The speed with which we serve our customers has become more critical than ever. We have bigger homes, better cars, want longer holidays, have greater demands, but there just isn't any more time. It stands to reason that if you can save people time you are providing better service; save people lots of time and you are creating five star service!

Let's take a few obvious examples to illustrate the point.

'To queue or not to queue?' Now that is a question. If we have to queue then at least we want to make the queuing experience as painless as possible. There's a strong argument that people don't mind queuing if they can see the queue moving. Probably true. There's a stronger argument that they would rather not queue at all.

'It's on its way.' When you order something, when do you want it? *Now!* If I order something over the telephone *I love to hear the words, 'It will be despatched today.' The time of 28 days for delivery is long gone.*

When I'm online I want web pages to appear instantly. Yet some sites still have you waiting while the wonderful (but time-consuming) flash animation appears to 'welcome you' to their site.

So what can be done to save those vital seconds, minutes, hours and days?

- **Step One** Put yourself in the customer's shoes and ask yourself, how long would I be prepared to wait for this? Then halve it – because most people aren't as patient as you. That's your goal – go on, stretch yourself.
- **Step Two** Look at your process, list every stage and ask, 'How can we save time with this?' A good way to do this is to use the system described in 'Spanners and Heroes' on page 217.
- **Step Three** Create a list and a plan, then add together every second, minute and hour. You may want to create a reward system as a target to save more time. Remember, every second counts. If you want to reduce queuing time and you think of six 10-second and

four 30-second ideas, you've reduced queuing time by three minutes. That could have a major impact.

- **Step Four** Check the balance to make sure you're not compromising quality for speed. This is a fine balance but many organisations (and particularly call centres) have been caught out recently by putting crazy targets on the length of time an agent can talk to a customer. Several found they had customers call back, taking up even more time because their query wasn't answered fully the first time.
- **Step Five** Test, test, test. You won't know until you test it so make sure you apply the ideas. I bet there are thousands of brilliant ideas still trapped on the turned-over pages of flip charts in offices around the country.
- **Step Six** Review every 90 days. Great ideas and best practice can easily fall by the wayside if you don't keep a regular check on how things are going.

35

Systemise routines – personalise exceptions

How long does it take to make a Mojito? The bar crew at the Blue Marlin Beach Club in Ibiza make them in less than 30 seconds – further up the beach it takes four minutes. At €14 each, which bar would you like to own?

I spent a very pleasant afternoon in August researching this chapter, observing how the bar crews worked, coped with exceptions, wowed their customers and made a tidy profit too.

Now flip it and think about it from the customer's point of view. August in Ibiza is a busy time, bars like the Blue Marlin Beach Club can attract well over 1,000 people. That's a lot of hungry, thirsty people who want to get their drinks and head back to the sun.

The staff know their most popular drink is a Mojito and they have a pretty good idea of how many they will sell. So well before the first customer arrives, they have a pre-opening Mojito preparation session. They make up hundreds of glasses with lime, brown sugar and mint. They know how long it will stay fresh and they time it perfectly. Next they have designed and set up the bar so everything is to hand. Finally everything else is stocked at one or two levels higher than requirement. They'll never run out of rum because the bar assistant replaces bottles when there are still four or five left. Ice is bought in pre-crushed and stored in an easy to reach central location. Each member of the bar team has their own utensils so there's no waiting around for a colleague to finish before they can make up your order. And finally (and most importantly) they look like they are having a ball!

Each part of the preparation and planning saves a chunk of time from a few seconds to a minute. And each saving of time is passed on to the customer.

Compare this with the beach bar just up the road. They start everything from scratch – which would be lovely if there were just you at the bar, chatting to the bartender as the sun sets. But there's not just you. It's five deep, everyone wants to be served and when you hear the person in front say, 'Four Mojitos please', there's an audible groan. They run out of rum, have to wait until their buddy has finished with the one 'muddler' and look completely exasperated.

The Blue Marlin has automated a routine and done it in such a way that the customer is delighted. Everyone likes to save time, even people chilling in Ibiza. The obvious question is, what can you do to automate your systems to make life easier for your customer?

Does your website have frequently asked questions?

Do you talk to each other about the most common challenges that crop up?

Do you observe how your competitors are running their systems and use the best of what they do and avoid the worst?

Could you shave a few seconds or minutes off your customer's wait time?

All of these ideas are simple, but not easy. They need time, effort and resources assigned to them. Then once you find your super system you can't relax – in fact you have to be prepared to deviate from it because it won't always be right and it won't always work.

The **last thing** a customer wants to hear is **why you can't** do something for them because you have **'a system'**.

Let's go back to the Blue Marlin and ask what they do when they get thrown an exception. Well, in the spirit of my five star study I ordered another couple of Mojitos (it's hard work doing all this research). This time I asked for one of them to have only a tiny amount of sugar but lots of mint. No problem for the crew who called it 'a special'. It took them just over a minute, no problem there but here's the best bit, when handing over the drinks the barman popped a different straw in to our high-mint low-sugar cocktail and with a smile said, 'That's your special one.'

Customers love routines to be speeded up, simplified and made more convenient, but never use this as a reason not to look out for the exception.

I hope the Blue Marlin continues to improve its Mojito-making system as I feel another research trip coming on!

36

Making the mundane marvellous

Which part of your customer interaction could be described as mundane? Completing forms, being left on hold, queuing? Even the most exciting businesses have moments that their customers find mundane. Let's start with a classic – queuing!

No one enjoys queuing; it's lost time and you can guarantee it creates stress. Here are my top five most stressful queues. See if you agree.

1 Having your call put on hold and not knowing how long you'll have to wait
2 Supermarket checkouts
3 The 'carriage queue' on the station platform – especially if you don't have a reserved seat
4 The petrol station queue towards midnight after the budget speech
5 Airport check-in for the homeward flight.

What can you do to make your customers' queuing experience more enjoyable?

1 Acknowledge the queue. If it's busy let people know. Give them a realistic expectation of what will be happening.
2 Give them something to do. Theme parks are the masters of this, often making the last 30 minutes' wait part of the experience.
3 Throw more resources at it. Is it beneath the MD to pick up a phone and take a customer service call?
4 Entertain. 'On hold' recordings are at best dull. How about a couple of jokes or some useful information?
5 And how about a sincere apology or acknowledgement that we've had to wait?

However, there's more to the mundane than just queuing. Filling in forms, refuelling your car, waiting rooms, watching a screen while your emails download. They're all a bit mundane and a bit boring. But what if you were to change that?

And it doesn't need to be a big budget flat-screen TV or special effects phenomena. Here's a simple one to get you started. I was refuelling my car the other day when I looked up and saw the attendant staring at me. I gave a smile and little wave and she just continued to stare back blankly.

I turned to see if there was anything more interesting happening over my shoulder, there wasn't. Now what could she have done?

Waved back and offered a cheery smile? Yup, that would be great three star service. Or what if she had waved at me first, or given a little thumbs-up, or simply acknowledged her customer? Let's make that four star. And for five star…? Well what would you do?

Here are a few more ideas to make the mundane marvellous. Some may be a perfect fit for you. Some may need a tweak. Some will be completely left-field but I'm sure they'll get you thinking.

Play music at bus stops.

Have a lucky car draw in a car park queue.

Have up-to-date, interesting, clean reading material in your reception.

Give your customer a chance to win their software free during a download.

Make a fun video with your staff to play while your customers wait.

Use a comedian (politically correct of course) for your 'on hold' recording.

Add a few encouraging comments on boring forms, 'Almost there', 'Last page', 'You did it!'.

Give staff training in non-verbal communication. Then use it!

Have a free Monday draw on a commuter train for season ticket holders.

Test, **test**, **test** is the trick.

Who knows if your idea will work unless you give it a go and learn from your successes and mistakes? Every time you test an idea you'll be raising your game and doing your bit towards making the mundane marvellous.

37

The distraction of dirt

Clean = Good

Spotless = Five Star

I'm writing this chapter on a train. It's big, seats hundreds of people and can travel at well over 100 miles an hour. I'm sitting in first class, which is a privilege I never take for granted, and your eyes (like mine) would water if you knew the cost of the tickets. There is a challenge though. It's a bit dirty. There are crumbs on the table from the last traveller, an unidentifiable stain on the window and the guard – sorry revenue protection officer – has dirty nails. The latter was pointed out by Christine, who spots nasty nail fungi at 50 paces.

It's not nice is it? I have a friend who got off a plane because his seat table was filthy. His thinking was, 'If that's how they treat the tables, how well do they look after the engines?'

And it gets worse. My friend Tom decided he didn't want to buy a £50,000 product because the salesman had dirty shoes.

There's a famous story that Michael Eisner, the then chief executive officer of the Walt Disney Corporation, was showing a group of international executives around the Magic Kingdom. He walked across the street, picked up a piece of litter and threw it in the nearest bin. When one of the group suggested he must have staff to do that he replied, 'The cleanliness of Disney is everyone's responsibility.'

I've heard the story so many times that there's a little piece of me thinks they must plant the litter for every tour so he can make that point! And the point is obvious – isn't it? Or is it?

The level of **cleanliness** around your product or service **has to be immaculate**.

Not good, not left to the next person, not 'it will get better tomorrow'. I'm talking about immaculate now.

And I'm talking about you as well as your environment. You get only one chance to make a first impression and customers make judgements with their eyes long before they have experienced the service.

Here's a list of eight things you must consider.

1. Take a good hard look in the mirror. Honestly, how do you look? And what could be better?
2. Check your immediate environment. How is the entrance to your building? I know it's 'not your job' to keep it clean but it's not Michael Eisner's either.
3. Nasty one. How are the toilets? If you don't think they are peachy then what do you think your customers will think of them?
4. Surfaces, signs and stationery. If they are dirty, your customers will subconsciously associate that with your offering.
5. Smells. If NCP car parks can do a brilliant job eliminating and replacing some of the challenging smells their venues are faced with (and I don't think I need to explain any more here) then what can you do with your environment?
6. Tidy. Make it easy for people to be neat by giving or getting the right storage and encouraging a bit of basic feng shui thinking.
7. Hands. Bitten or dirty nails, picked skin and nicotine stains all add up to a poor personal service.
8. Body smell. Bad breath, BO and festering feet. If you think you have a bit of a body odour issue then the chances are you have! If you're worried that you may have bad breath you will be the last person to spot it! And if you think your feet smell a little, it's more than likely they're lifting! You need an honest Joe who'll tell you (and you can tell them) if there's a nasal nuisance to sort out.

38

Send cards

Did you know that the average person receives fewer than four cards each year for their birthday? Did you know that the average person receives fewer than 10 cards a year for all other occasions? Did you know that the average customer receives *no cards* from their suppliers – ever.

We post them bills, we send them contracts, if we do despatch cards they tend to be mass-produced corporate Christmas cards with a pre-printed message. We rarely send customers cards for anything else. Yet sending cards is a great way to show people that you really care. It's a small action with a massive value. They can be sent to internal and external customers with equally brilliant results.

I once worked with a division of HSBC bank on a project, part of which involved a campaign to send every single member of staff a thank-you card to show how much they were appreciated. When you work for one of the biggest companies in the world it's easy to forget what an important part you play. There are thousands of staff in the division so we needed to devise a method to produce a personalised card. Here's how we did it.

We had a branded card designed and produced (blank inside leaving plenty of space to write a personalised message). The divisional general

manager personally bought all the cards for 10p each and donated the money to the HSBC charity fund. She kept a batch of cards for her own use, then sold boxes of cards to her area directors for 20p each and again all the money collected was donated to the charity. They then kept their own batch and sold the rest to their managers and team leaders. On the back of each card we had a statement printed: 'Sold up to three times for charity.'

Then in the next 90 days the leaders had to find a way to thank their staff, sending them a personal card with a handwritten message. Everyone knew they would receive a card, but they didn't know what for or when. After 90 days there were cards everywhere and everyone had a personal 'feel good' moment (in fact most people kept their cards for months afterwards).

The real people to benefit were the customers. Because happy staff, who are recognised for the good work they do, are better at customer service.

Note to managers – **treat your staff the way** you want them to **treat your customers**.

So what could you send cards to your customers for?

- Birthdays
- Paying a bill on time
- Being patient
- For a referral
- For a great meeting
- For an anniversary
- For Christmas
- To celebrate their success
- As part of an apology

Bonus Bit

Here's a tip on how to make the card even more special.

As well as handwriting the inside, handwrite the envelope and put a live stamp on it (don't use the franking machine).

Think about it: if you get a card in the post with a handwritten envelope and a live stamp among a whole heap of other mail, which are you most excited about opening?

Other ideas around the card theme

Flip chart thanks

If you have a small team, take a bunch of coloured pens and a flip chart pad. Do a page of 'thank yous' using colour, space and imagination for each person in your team. Stay behind late and stick a personal page on an individual's desk, their workstation or on the wall. Arrive slightly later the next morning.

Blog thanks

As blogging becomes an increasingly popular way to let people know your thoughts online, perhaps a blog thank you may be suitable with a link to the subject's website.

Send a letter

With the advent of email we don't send as many letters as we used to. If you don't like the idea of a card, a nice letter could make someone's day. When writing a letter, it's a nice touch to handwrite the 'Dear' at the start and it's a must to handwrite the 'Yours sincerely' at the bottom.

Photo thanks

After a successful meeting we'll often get back to our office and write a thank-you message on a flip chart, stand next to the message and have a digital photograph taken. We email the picture to the person we've just met with – they love it!

39

Designing fantastic service

Have you ever held an Apple iPod? Have you ever used one? This piece of technology has become a design icon in just a few years but it didn't happen by accident. Teams of people spent thousands of hours making the iPod simple to use and stunning to look at.

Take the car manufacturer Lexus. The people there asked prestige car drivers all over the world what they liked about their BMWs, Mercedes, Jaguars, etc. They then took all the ideas, removed duplications, then *added* over 500 improvements and set about designing what they described as the world's finest luxury car.

The finest designs often take masses of hard work to make them appear simple. Working with hundreds of organisations, I've seen various levels of commitment to designing wonderful five star service. Many get very excited about starting a process with 'off-site' meetings, sizzle sessions and then wheel in the external consultants who happily charge big money to ask (with an ever so serious face), 'Tell me what you think ...' That's an expensive route. It's fine if you want to spend a small fortune, but how do you do it on a one star budget?

I met the fashion designer Wayne Hemmingway, at the time my daughter had her heart set on being a designer, so I asked him whether he had any advice for her. 'Does she make her own clothes?' was his first question. He was asking whether she was just into the image and the glamour or whether she was prepared to work hard and really create something beyond a drawing. His next bit of advice was for her to make me a shirt and clothes she could sell to her friends. All this at 13 years of age? He's right.

Our discussion led me to a conclusion. If you really want to get into a prestigious fashion school, and you've been making and selling clothes for five years, you're going to have a pretty big tick on your CV.

'The secret isn't in the knowing, it's in the doing.' It's something I say a lot. It's the same with **designing brilliant service.**

Think about it. You can have all the theory in the world but until you actually test it out you have no idea whether it's really going to work. When Thomas Edison invented the incandescent electric light bulb he **tested** over 5,000 ideas before he found the one that worked. If he'd only thought about 5,000 ideas, we could have been sitting with candles for hundreds more years.

So how do you design world-class service? Here are seven keys and a simple process to get you started.

1. Know what you want. If you don't have clear objectives, how will you know whether you've got there?
2. Get people involved. Do your own version of a public consultation and listen carefully to what everybody has to say. You don't have to take all the advice on board. And remember, people will soon stop sharing if you constantly interrupt.
3. Create two piles of cards, 'all the things that can go wrong' and 'all the things that must go right'.
4. Share the cards around your team and allow people an opportunity to build on the rights and eliminate the wrongs.
5. Lay out the cards showing the 'flow' of service. This should show the links, the strong points, the potential weak links, etc.
6. Run the sequence forwards *and* backwards asking these three questions:
   - **i** How is the customer feeling?
   - **ii** How are we feeling?
   - **iii** What can we do better or change?
7. Once you have your 'service route' planned, ensure individuals have responsibility and timescales for making it happen, then review the system on a regular basis.

By making your **thinking three-dimensional** and running the sequences forwards and backwards you will **pick up** many of the **potential challenges**.

Bonus Bit

When you run your service sequence, have some gold stars. Give a star to each part of the sequence that would outshine your customer's expectations. The goal is to get a minimum of five stars in every sequence.

Creativity gives better service

Simple creativity is sometimes staring us right in the face. It's one of those things that you look at and you think, 'that's brilliant'. Usually the greatest ideas are classic one star budget ones.

Some brilliant examples

One of my wife's favourite movies is *Pretty Woman*. If you are one of the three people in the world who hasn't seen it, you'll need to know much of it is set in a suite in the Regent Beverly Wilshire Hotel in Beverly Hills, Los Angeles. The Beverly Wilshire has been voted one of the best hotels in the world and it is also one of the most creative, mixing elegant style with some really cool ideas.

Having watched the movie for the hundredth time I promised my family we would visit the *Pretty Woman* hotel and, during a trip to America, we found an opportunity to visit the West Coast for a few days. Only one slight challenge. If you want to relive the *Pretty Woman* experience and take a suite at the Beverly Wilshire Hotel, ideally you'd arrive as just a couple. But we love travelling with our kids so what do you do?

Well, the Beverly Wilshire knew that lots of people wanted the *Pretty Woman* experience but they also travel with their kids. So the hotel catered for this with a package called 'Not just the two of us'. It was great! Mum and Dad got the fabulous suite with the champagne and strawberries, and the kids got an adjoining room. But it was even better than that. They filled the kids' room with sweets, 'soda' and ice cream, gave them free movies and put a bunch of games in there so they didn't want to leave.

Then there are the outings. The concierge asked our kids what they wanted to do first and then asked us, the parents. He then seemed to beautifully mix everyone's requests in a perfect schedule. For example, we wanted to do a 'stars' homes' tour. The concierge could have suggested this one or that one but instead he asked us to make a list of our favourite stars. Then he looked at the various tours to see which one fitted best. Typically, no tours featured our requirements so he arranged a private tour. The concierge told the tour company who we wanted to see so they could work out a route in advance, giving us more time for the tour.

On the flight home we were talking about our favourite parts of the trip and everyone's lists were based around the experiences more than the rooms, food, etc.

This doesn't apply only to top hotels, however. Notice how the actual 'experiences' aren't cases of just throwing money around – it's about anticipating what would make your customers' lives easier, being flexible and above all being creative with your solutions.

Here's an example in a less glamorous setting. Cheryl Black is Customer Services Director with O_2 but she has a track record of helping organisations, such as Scottish Water and NTL, create wonderful customer service.

When NTL first enabled people to subscribe to individual movies via its cable connections it experienced some teething problems. Customers would call up and say they had subscribed to the movie and then it didn't arrive. Of course they were entitled to a refund, and as an act of goodwill NTL would offer a free movie on another night. That was fine, but if the customer's Saturday night had been ruined because the movie wasn't available, they weren't really thrilled with this. NTL needed to take it to the next level.

Here's what they did and it made a significant difference. If there was a challenge with the delivery of a movie and a customer called to complain, they would resolve the problem and then ask, 'Do you prefer Chinese or Indian food?' After their response, the customer service adviser would say, 'The next time you watch a movie, let us pay for your takeaway. You order it and we'll put a credit on your account now to pay for it.'

What a brilliant solution. Recognising that what the customer was really upset about wasn't so much the missing movie, more a ruined Saturday night in, led them to devise a more appropriate response: not just a credit on a bill, but an even better Saturday night in next week! It's what they called 'Making the emotional connection'. When you think about it, it's a win win and it achieves several things:

1. The problem is solved but they go the extra mile.
2. They add value in more than a financial way.
3. They encourage you to watch more movies.
4. They make you feel good – very good.

Another example comes from a low-budget local Spanish bar. During one of our *5 Star Service* workshops we heard this example from one of the participants and everyone agreed it was one of the best we'd heard.

Danny and his family were on holiday in Spain and on their first night they went to a local bar where they had karaoke. Danny got up to sing and wowed the crowd. The owner commented on his enthusiasm and bought him a drink. The next night he was there again and the following night. That night he told the manager he was going to visit some other bars but he would be back. On their last night, they were walking to the beach and on a board outside the bar they saw a sign that read,

'Tonight, for one night only – Danny.'

Danny was delighted. Not only did he take his family along, but because of the gregarious type of person he is he took 25 other 'new' friends he'd made too.

Actions

Here are five ideas to get your creative juices flowing:

1. Think like a customer – literally. If you can, see how far you can get through your processes as a customer. Ask yourself, what could you do better?
2. Have a brainstorm – but do it properly. You know the rules by now for brainstorming: every idea counts, no negative comments, etc.
3. Ask how nature would deal with it. Nature has a great way of adapting to situations. How would nature deal with your service to make it even better?
4. Free up creativity in others. If you are a manager, allow your teams to think of creative ways to help your customer service experience.
5. Ask, 'What if?' questions. It doesn't mean you'll do it but it's a great way to get you thinking differently. Here are a few to get you started. What if:

   we gave away our products for free?

   our best customer was going to leave?

   I had all the time in the world?

   we doubled our budget?

   a member of the royal family was coming to visit?

   we were entering a competition for a customer service award?

41

Suppliers are customers too

It's all about price. Isn't it? Well yes and no. We all want a good price but I've yet to meet anyone who would admit to wanting to sacrifice great service for a little extra discount. But the fact is people do. And they do it with their suppliers financially, emotionally and habitually.

Remember this.

Customers can be defined as **anyone** you have an **interaction** with.

This means your suppliers are your customers too and how you treat them will have a major impact on your other customers – the ones who pay your bills.

After many years in business I'm happy to say that I believe we have found a group of brilliant suppliers who are committed to great service. I don't know how good they are for their other clients but I do know they are brilliant for us. I know that because I've worked hard to help them be brilliant suppliers. Here are a few of the things we've done in the past and do now.

- Invited them to participate in our events.
- Thanked them in writing and let individuals' bosses know how much we appreciate their great service.
- Referred them to other companies.
- Invited them to our Christmas party.
- Sent them copies of my books.
- Written about them in newsletters.
- Created links to their businesses on my website.
- Taken them out for lunch to talk about challenges.
- Shared with them future plans and listened to their advice.
- Paid our bills on time.

Getting the best out of your suppliers is exactly the same as getting the very best out of your most loyal staff, and if you want to create a brilliant five star experience then having brilliant five star suppliers is a must.

I really struggle with organisations who bring in companies or new staff whose sole aim is to reduce costs. What if you employed a member of staff whose sole role was to improve the relationships, loyalty and service from your suppliers? Do you think you might find some cost savings in there too?

And does it work? I really do believe so. I know that if we have an IT problem Norman or Neil from Datawright will dive into action and go the extra mile to get it fixed, often working overtime or on problems at home. I'm certain that if our website develops a fault, Alan, or a member of his team at Tricycle Media, will fix it faster than our agreed response time. And it's not just the little businesses that can do this. We have a fantastic relationship with HSBC and I'll continue to write for Pearson, not because they are the biggest publishers in the world but because they have built a relationship with me that enables me to get my message to my customers.

Here's a simple way to check then improve your relationships with your suppliers.

1. Create a list – here's a clue, they are probably sending you invoices so just print off your creditors.
2. Consider what you want from each supplier, ask your colleagues, then write it down.
3. Let them know. Meet up, pick up the phone, have a suppliers event.
4. Keep them informed. Especially when they get it right. Thank-you cards, personal cards, small gifts, etc have a massive impact.
5. Pay your bills on time and, if you can't, communicate why you can't and when you will.
6. If you get a better offer from another supplier, talk to your existing one before switching. There may be a very good reason why they charge a little more. Who knows, they may even price match.

42

Service PR

What are you known for? Could you win a Customer Service Award?

When you completed the Service Star™ in Chapter 1 you'll have noticed one of the potential areas where you can score high marks is service PR. In simple terms this is what people are saying about you, often in the media but more so these days online, between customers and even within your own staff.

Some organisations spend fortunes with public relations companies developing gimmicks and ideas to spread the word about their amazing (usually their words) service. Even with all that spend they still get poor results.

Others spend huge amounts trying to keep their lousy service reputation out of the public domain. You can imagine the scenario at the board meeting, budgets are being carved up and the big cheese is terrified in case he loses his 'defence PR budget'.

Then there are some who have plenty to shout about, use just the right amount of PR to fan the flames but leave the majority of their positive PR to their customers; loyal customers who are happy to rave about their experiences and defend the company wherever necessary.

So how do you go about creating fabulous service PR? Here are seven steps to get you started.

- **Step 1** (And this should be blatantly obvious) create fabulous five star service. I had a client who wanted to win a national customer service award, and asked me how I thought he should go about it. After I'd explained what I thought was necessary, he asked, 'Is there an easier way?'
- **Step 2** Make sure everyone in your organisation knows you want to create fabulous service PR. Again this seems glaringly obvious but you'll be amazed how many people want to keep their service sensation a closely guarded secret. Encourage your team to share successes, learning experiences and solutions to customers' challenges.
- **Step 3** Document what you do. Having a great story you can tell is good. Having it documented for others to read is brilliant.

- **Step 4** Share the success! Write a newsletter. Send out news releases. Employ a PR company. Fill your website with how wonderful you are.
- **Step 4.5** Get over the fact that step four didn't achieve much! Brilliant service PR takes a bit of time and needs something far more powerful than you shouting about how good you are. That's right, it needs your customers to shout.
- **Step 5** Give your customers something to shout about. In his brilliant and free eBook *Flipping the Funnel*, author Seth Godin talks about giving your customers a voice to promote you. Imagine what your service PR would be like if you could show your customers how to shout about you.
- **Step 6** Get recognised. Enter your organisation for a national customer service or WOW award. You might just win. Or write to me about what you've achieved. I'm always happy to share examples of service success if I think it will benefit my readers.
- **Step 7** Be sure you can walk the talk. Promoting yourself as a five star service organisation means you can't hide behind inexperience, downturn, poor training, company policy or any of the other excuses you may have clung on to in the past.

Service PR can't be bought – or if it can it doesn't last.

Service PR comes from being able to genuinely **celebrate your service success** and have your customers do the same.

43

The blind spot

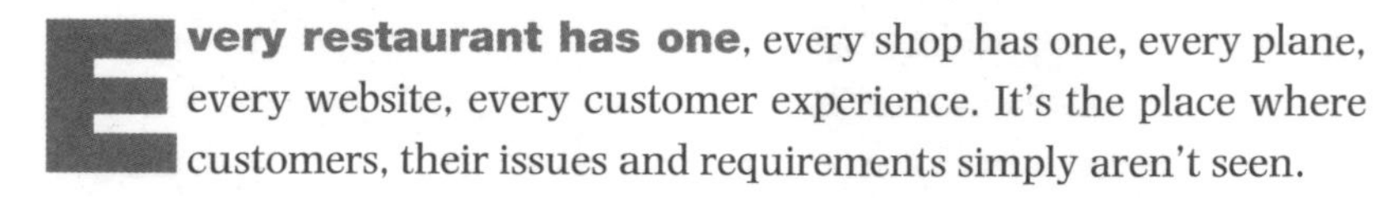

Every restaurant has one, every shop has one, every plane, every website, every customer experience. It's the place where customers, their issues and requirements simply aren't seen.

One of my favourite restaurants is Caffe Vivo in Newcastle. It's close to home, the food is amazing and the service is magic. We've been many times with family, friends and on business. We have never been disappointed.

Then one day a funny thing happened. We sat down for a meal and browsed the menu. With our mouths watering we were ready to order. And we waited ... and waited ... and, because I'm not very good at waiting, I did a slightly embarrassing wave before our order was taken.

The starters were sublime, the mains magnificent but when we were ready for more drinks the problem recurred. And I ended up waving.

The challenge became obvious – we were in the 'blind spot'. A part of the restaurant where the diners were less visible. Because we know how brilliant they are at Vivo it didn't affect our night – they have made enough

deposits in our emotional bank account to make a small withdrawal. However, the real issue was they didn't know they had an issue until we pointed it out.

That's the problem with the blind spot – it's not that you don't want to fix it, it's that you don't know it's there to fix. And blind spots come up everywhere. It could be:

- the navigation of your website
- the wording of a question on your application form
- the layout of a shelf
- a misleading pricing policy
- a glitch in your software
- being unaware of the attitude of someone on your team.

So how do you identify your blind spot and, when you find it, what do you do?

Here's where your customers would love to help, but only if you ask them properly. If you say, 'How was everything today?' You'll get 'fine' as the response. Even when it wasn't.

So how about asking a better question after the 'fine' response? Something like, 'Thank you, but if there was one thing we could do better what would it be?'

You can find out more about this in Chapter 7 'Beware the silent customer'.

Another way to **find your blind spot** is to test, test, test; watch, watch, watch; act and review.

Most people do a little bit of testing a tiny amount of watching and little or no acting or reviewing. You're not most people.

Computer game designers employ testers who play their games for hours at a time, running outrageous scenarios and monitoring the results. If you were to really test your system would it hold up? Test doing things at different times, test where your staff are positioned, test speeds, layouts, eye levels, in fact, test everything.

Watching your own business is often considered a luxury, especially for business owners. We suffer from the classic leadership conundrum of spending too much time working 'in' the business rather than working 'on' our business. If you owned a pub, how comfortable would you be just to observe how your customers act, notice how often they drink, listen to what they ask for and be aware of their needs? I'm sure you'd be very comfortable with that, but the chances are you don't own a pub, so what's your equivalent?

And now for the easy part – take some action.

Big Buyer is watching you

I don't know about you but I'm a compulsive eavesdropper. I can't help it! I love to listen in to other people's conversations on the train. I'm fascinated with what's going on behind the scenes in restaurants, shops and hotels and my absolute favourite is earwigging when two members of staff are talking about their company. Isn't yours?

The easy answer is to claim that you would never do such a thing and you respect a person's privacy. But can you honestly tell me that if you were in a shop and you overheard two members of staff criticising their company, you wouldn't hang around for a listen? And then you'd decide whether or not you wanted to hand over your hard-earned cash.

Yes Big Buyer is watching, listening and reading about you.

Out of sight, out of … mind? Customers are nosy; they love to be in the know and can't wait to catch you out. A dirty backroom or disorganised office will create instant doubt in a customer's mind.

You get only **one chance** to make a **first impression**.

I remember when I first started my business. I had nothing but my enthusiasm. I even had to borrow an old car (with no brakes) to get to appointments. During my first week of trading I set a goal to see 20 new customers. I arrived outside the offices of my first prospect and skipped in to the meeting. I pitched and promised to make them super successful and explained in detail exactly how I'd do that.

After the meeting, as I was getting back into the car (the one with no brakes), I looked up at the office window and was horrified to see the person I'd met with peering out of the window. Through a sarcastic smile he mouthed to me 'nice car' as he pointed at my borrowed wreck.

What kind of image was I portraying? One minute claiming I could make them more successful and minutes later looking like I couldn't even do that for myself! Big Buyer is watching.

If you have any area of your business that you wouldn't want your customers to see then change it, and create an environment you can be proud of.

Customers are listening too. Clients find it rude and disrespectful when staff are having conversations among themselves instead of dealing with their customers' needs. Big Buyer is listening.

And it's never been easier for customers to have a voice through blogs, review websites, instant messaging, social media, etc. You can't stop it but you can learn from it.

If you're embarrassed by what people say about your organisation online then what are you going to do about it? Whose fault is it when things get out of hand?

Would you want to board a flight if the first image you found when doing a Google search of the potential airline was of drunken staff on a boozy night out? I'd think twice too.

Your customers are watching you. And more closely than you think. Your conversations are overheard, your Facebook pictures are probed and your 'secrets' are common knowledge.

But it's not all bad news. Because customers are more savvy than ever, here are some things you can do to show that you care.

- Search Google with your company name followed by 'customer service'; get past the corporate fluff and read what people are really saying.
- Check your 'staff only' areas. Would you be happy for your customers to see them? If you have a kitchen would you be proud to show it off?
- If your staff are going to talk about their company, make sure it's positive. What do you want your customers to hear?
- If you get negative feedback online don't hide, respond to it. See Chapter 6 'Embracing new technology' for more information on this.

45

Secret shopper

If you can afford it (and maybe even if you can't), I would recommend you consider using a secret shopping service to help you raise your game. Good ones help you to:

- remove yourself from the process
- identify what's wrong
- recognise what's right
- test your systems
- improve your sales
- help create or carry out training with your staff.

The basic idea is very simple. You brief a secret shopping company with what you would like to find out. They shop, eat, call, etc your organisation then give you feedback against the criteria you discussed in the initial briefing. There, it's that simple.

You need to have a very clear idea of what you want from a secret shopper experience and you have to find someone who's on your wavelength.

The Secret Service (no not that one) is a secret shopping company that specialises in making the whole experience enjoyable, educational and profitable. Its founder Linda Eastwood explained to me that, by getting the right secret shopper company to help your business, you should increase your sales, improve staff loyalty and find out as much about what you're doing right as what you're doing wrong.

When I met with her she was keen to dismiss a few myths about secret shopping. Here's a selection.

Myth One Don't tell your staff you are using a secret shopping company, that way you'll get feedback warts and all.

Do tell your staff! Why shouldn't they know that you're using secret shoppers? Don't you trust them? Because once they find out you're 'sending in the spies' they won't trust you.

Myth Two If you let your staff know won't they start to improve their level of service before your secret shopper arrives, and won't that defeat the object?

Now your customer service is improving already – get it?

Myth Three It's good to use a secret shopping company because if someone else catches them doing something wrong, then I'm not responsible.

But this is about everyone's reasonability. It may be that you have to face some home truths about your levels of training, systems and investment in your people. Are you ready for that?

Myth Four Won't it demoralise my staff?

That very much depends on how you treat the experience. Do it in a sneaky underhand way that focuses on what's wrong and yes you may demoralise them. Make it a fun experience that rewards what's right and staff will hope that they are being secretly shopped.

Myth Five We have only a small business, so I'll just ask my friends or family to help.

Often people with small businesses make the huge mistake of asking friends and family to secret shop for them. Our friends then think they are doing us a huge favour by criticising everything from the colour of the wallpaper to the number of rings before someone picked up the phone. The result is a disillusioned owner with friends who have the impression that you have a rubbish business.

If you can't use a professional company, then how about using a member of a trade group you belong to? You could return the favour for them but remember to keep it positive.

Secret shopping is as much about observing people **doing things right** as it is about catching them doing things wrong.

46

Special requirements

My wife is lactose intolerant. Basically this means that if she eats anything with cow's milk, cream, cheese, yoghurt, etc she's violently ill for several days.

Unfortunately, if you don't know anyone who suffers from this intolerance or you aren't aware of the effects after an accidental mouthful of the dreadful dairy, then you may just think she has a fussy food fad or is testing out the latest Hollywood diet. She's not.

So you can imagine what it's like for us ordering in a restaurant. 'May I just check, does that have any dairy in it, please? I'm lactose intolerant and can't have any dairy products,' asks Christine.

'I don't think so,' replies the waiter and the order is placed. Our dinner arrives. Christine toys with her potato, then asks the waiter, 'Are sure there's no dairy in this?'

The waiter then takes a huffy strop, goes back into the kitchen, struts back to the table and announces, 'Chef says there's a tiny bit of cream in the potato.'

That's like saying to a vegetarian, 'There's only a small amount of beef in your meat-free mushroom risotto.' A 'tiny bit' of cream once gave Christine three days of agony.

Here's the flip side. The Rubens Hotel in London knew Christine had a challenge with lactose (a sister hotel in their group let them know – brilliant!). We were staying only one night but became instant fans when we checked in to our room and found two room service menus. The standard one and a second one, marked up by the chef, with all the lovely food Christine could have. We were going to eat out that night but, guess what? Yes, we stayed in and ordered room service.

When your customer has a special requirement you have a choice. You can moan about it, try to placate them and give them your second best option.

Or you can **wow them** by providing a brilliant personal

service, show that you really **understand** their needs and **create a fan** who will talk about how **amazing you are** to anyone with ears!

And it isn't just food. Some people may want to meet you early in the morning, others late at night. For some it might be religious views, which you may not understand but are fundamental to their way of life. Many people have disabilities, some very slight, but they still deserve your consideration.

And if you're not sure what a person's requirements are or how you can help them, then ask. I've yet to meet a person who has a special requirement who minds being asked about it. No matter how obvious it seems.

Finally use a 'heads up' approach (see Chapter 3) and be aware of the less obvious times when someone may need help. Large-print price lists, different language answers to commonly asked questions, a brief understanding of other cultures and an awareness of your environment and how it may affect others can sometimes be the small thing that makes a massive difference.

47
Sell
me a
solution

I like Sky. I subscribe to sports, movies, documentaries, high-definition, multi-room, everything really. So when it came to searching for a new broadband provider I looked at their offering of lightning speed and low cost (still waiting for the lightning speed here in rural Northumberland by the way) and ordered my new service.

The order process was a chore, but as a loyal customer I forgave them, just so long as they fulfilled their promise of the new product by Christmas. Then we had some delays, which I was able to overlook – as a loyal customer. Then the router etc arrived on Christmas Eve and only one of our computers worked (luckily mine, unluckily not the kids'). Can you imagine the horror of being a teenager without Facebook or MSN Messenger over Christmas?

Then the fun really started. For two weeks I had almost daily calls with a wonderful array of technical people. All were polite, helpful and patient. However, none of them managed to connect our laptops to the internet. Then, after various levels of 'escalation' I was eventually connected to a techie of such knowledge and standing I wouldn't have been surprised to hear that he had invented the internet!

His first question was, 'Are you using a wireless USB Sky dongle in your laptop?' I didn't even know there was such a thing as a Sky dongle, so I asked him where you got them from. He informed me that, 'Sky sell them, they are £20 each and can be delivered via a "next day" service.' Problem solved.

Here's what puzzled me. Why had I spent an average of two hours a day, three or four days a week for the past two weeks talking to the lovely patient people at Sky technical support, resetting IP addresses and goodness know what else, when the problem could have been fixed for £20?

The solution was easy.

But why wasn't I offered this solution by any of the other many, many people I talked to? Because it would mean they had to ask for a sale, and many people have a limiting belief that asking for a sale doesn't always sit well with great customer service.

There are of course two sides to this argument, which can be answered with one simple question.

Will my customer be **better served** and **feel valued** if they have to **make a purchase** to find their **solution**?

Approached in the right way, and with the needs of your customer put first, most people are delighted to pay for a solution, if it offers real value. Here are some examples:

> You'd like 'at seat' service, wider seats and a comfy quiet journey? Then you'd buy a first-class ticket.
>
> You want to come in from work and find your ironing done and the house immaculate? Then you'd employ a housekeeper.
>
> You want your computer to work brilliantly, never crash and be able to store hours of video and millions of photographs? Then you'd buy one with a huge memory and a better processor.

Five star service **doesn't mean you can't charge more** for providing better value.

In fact, selling the solution may be the best way to create a brilliant customer experience.

On the next page you'll find a simple way to help you remember

The five star service time/Money matrix

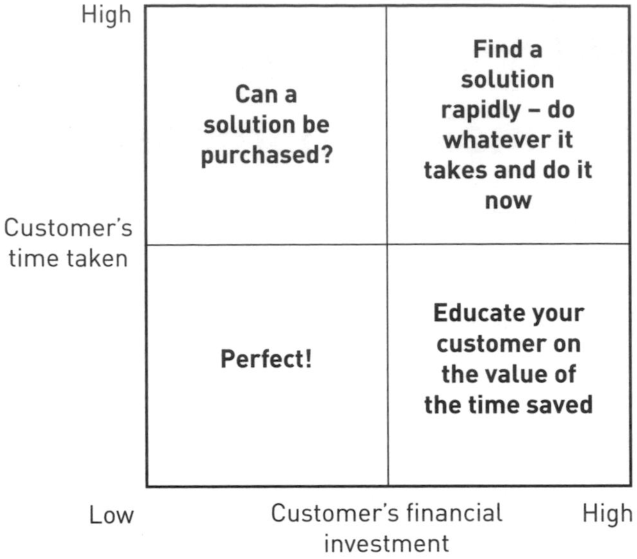

High	**Can a solution be purchased?**	**Find a solution rapidly – do whatever it takes and do it now**
Customer's time taken	**Perfect!**	**Educate your customer on the value of the time saved**
Low	Customer's financial investment	High

Low time/low investment The perfect place to be. Everyone is happy.

High time/low investment If the solution can be purchased you must let your customer know. They have to be given the choice.

Low time/high investment It's important to educate your customers on what they are investing in. Share with them the results they can expect from their investment.

High time/high investment OK, not the best place to end up, but now you are here take massive action to fix it. What extra value can you give to your customer?

48

Take it from the top

You've worked hard, you're on the way up and well done – you deserve it. However, this meteoric rise is taking you away from your customers. Should you worry? Should you do something? If so, what?

Easy. Take a leaf out of the book of Captain Denny Flanagan. I was introduced to Capt. Denny by Anthony Williams with this instruction, 'Michael, if you are writing about customer service you *must* feature Denny Flanagan.' Capt. Denny is a highly regarded, long-time pilot with United Airlines with so many flights under his belt he could choose to take it easy and leave the cabin crew and dispatchers to do the direct customer service work. But he doesn't. In fact he's a brilliant example to all of his team of how to look after customers at a grass-roots level.

I had to find out more about him, so I tracked him down and asked if he would like to be part of the mission! Here are a few extracts of what he sent back to me.

> Dear Mr Heppell,
>
> Thank you for reaching out to me. I looked you up on Google and was quite impressed with your accomplishments and I will take you up on your offer to help spread the word about just treating people NICE.
>
> If employees only realised that their paycheck comes from the customers walking thru our doors then they would be more positive coming to work in the morning. I'm working from the bottom up but the effect is still contagious. There is a genuine dire need to satisfy our customers by some employees and we have developed a bond and the group is growing.
>
> In the service business the recipe for success is quite easy. Anticipate your customers' needs and exceed their expectations. I have a few work philosophies and they have proved effective over the years:
>
> Treat each customer as if it is their first flight and have no expectations ... I lead by example and this helps motivate the crew to do a better job. When they (the other staff) see me stow bags, assist moms with strollers and answer a question as if it

is the first time I heard it they are brought back to their new hire days.

It is easier to keep the customers you have than to find new ones . . . United has a devoted sales team to find new customers and it is time-consuming and expensive but necessary. My job is somewhat easier and less expensive and that is to provide a safe and customer-oriented service. If I do my job then the folks in the sales department will have less pressure on themselves.

I have an array of ideas to connect with my customers and I keep them in my mental tool box. I'm constantly adding ideas and the tool box never gets heavier but my job just gets easier.

For years I have written notes to everyone in first class, including employees. Customers I thank for their business and employees for their personal and professional effort because it makes a difference for all of us at United Airlines. I would also randomly send 20 plus cards to customers in coach.

I loved those ideas from Denny but my favourite was this:

On Fridays and Mondays the aircraft is predominately filled with business travellers who normally get the aisle or window seats. Many passengers have to sit in the middle and it is quite uncomfortable. The window and aisle customers claim the armrest and for the next few hours there is the strategic movement of arms to claim the coveted armrest. Your seat mates may not even talk to you because that will reveal their nice side and then they will have to offer you their territorial claim.

My latest tool/idea is an attempt to make the middle seat psychologically and physically larger for my customers. For about two years I have written notes to the middle-seat customers on Mondays and Fridays. When the flight attendant leans over and says, 'Mr Heppell, I have a note for you from the captain', two things happen. Emotionally your seat just got bigger because you were recognised by the captain. Physically your seat became bigger because your seat mates move a bit to the right and left because now they want to know all about you and be

your best friend. And now you have both armrests for the rest of the flight and they are talking to you.

Wow! Who wants to fly with Capt. Denny? Me! Me! Me! I know you don't fly planes (or you might) but what's your equivalent of making your customers feel amazing, important and loved?

Oh yes, there is one more thing Capt. Denny does that really blew my mind. There's not enough space here so I'll add it to the free bonus chapters you can receive on page 236.

Wee Wow

Here's how he finished his letter to me:

Michael, statistics reveal that for every compliment or complaint received that there are 100 others thinking of doing the same thing. Receiving your note from another continent brings joy to my heart that my positive efforts are spreading. Thank You.

Capt. Denny

330-***-**** (My cell, in case you have any issues with United that I can assist with.)

49 Hills and valleys

s you study these techniques you will notice that you go through several distinct stages of learning. There are five and I call them 'hills and valleys'.

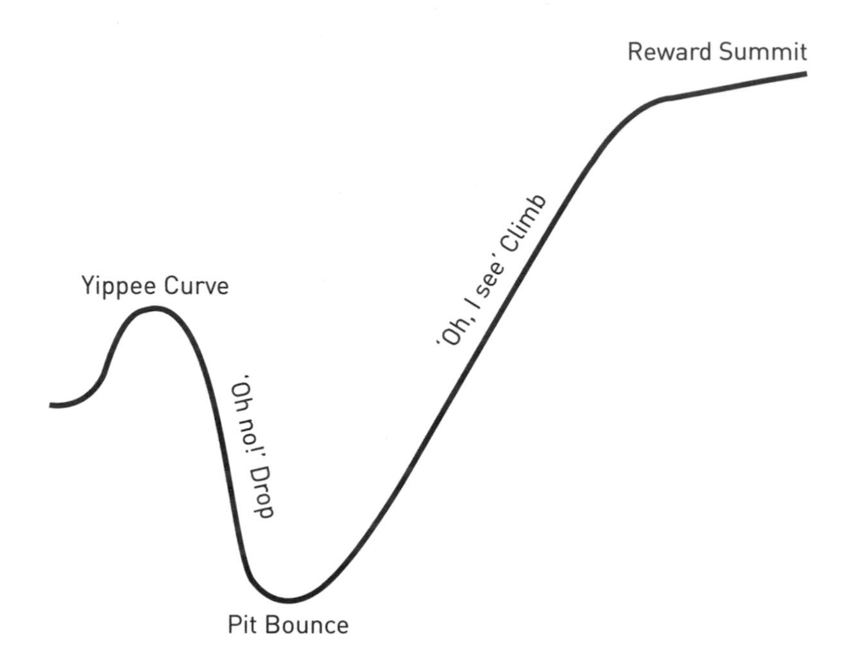

Yippee Curve

The first is the 'Yippee Curve'. The Yippee Curve is a great place to be because you start to use your new ideas and you get some amazingly quick results. In fact, you'll be so excited about using the ideas from *5 Star Service* that you won't even notice if things go wrong. And if you're part of a team of people who are all using the techniques then the Yippee Curve will be even longer.

'Oh no!' Drop

The next stage is what I call the 'Oh no!' Drop. 'Oh no!' Drop is an interesting place to be because, even when you are using your tools and techniques that you've learned from this book, you still seem to get things wrong. This can be very frustrating but it's not unlike learning any new skill and being shown a new way to approach an old habit.

I once took some golf lessons and for the first few rounds afterwards my game hadn't improved at all. In fact, I was concentrating so much on the new tools and techniques that I'd learned from the professional that my game became worse. But it wasn't long before the investment in the new ideas started to pay off and I was able to hit the ball straight and further than ever before.

Pit Bounce

The next place is what I call 'Pit Bounce'. It may seem daunting but it's actually a great place to be – because as you get to the lowest point there's going to be a big bounce which will catapult you into the next stage. It's all about sticking with it. I see so many people quit because they've got to the pit and haven't realised that there's a bounce that's going to take them up, up, up. It can be a confusing place to be, but that's OK because the next stage is what I call the 'Oh, I see' Climb.

'Oh, I see' Climb

The 'Oh, I see' Climb is where it all starts to make sense and your hard work really begins to pay off. The techniques become easier and easier and five star service becomes a way of life. You really begin to fully understand the ideas and make them work even better for you and by doing so you start to master the techniques.

Reward Summit

The final level is Reward Summit. This is where it all starts to really happen and is the best place to be because you begin to reap the benefits offered by the ideas in this book and start mastering them.

Realising that you're going to progress through these five stages is an important part of implementing these ideas. In fact, if you don't go through all five stages I really believe you aren't going to get the best out of the learning process.

So when you get frustrated and some of the ideas appear not to be working as well as you had expected, recognise that you're probably going through an 'Oh no!' Drop, adjust your behaviour, take some positive action and celebrate the bounce.

Actions

Five things to do as you go through each stage:

Yippee Curve	Enjoy it! Make the most of every new tool and test them out.
'Oh no!' Drop	Recognise it and ensure you have people around you to support it – never ever quit.
Pit Bounce	Don't dwell. Rewrite your goals and make a decision to aim even higher.
'Oh, I see' Climb	Document your learning. As you realise how well you are doing, it becomes easier to apply the ideas and sometimes you forget how you got here.
Reward Summit	Teach other people what you have learned. Give something back.

50

Spanners and Heroes

**ere's a great game** you can play as a team to learn about and improve your customer service. You'll need:

- a pile of A6 cards
- pens
- some coloured wool
- some pictures of spanners and superheroes.

I love the expression 'spanner in the works'. It conjures up a very visual representation of how customer service can go so wrong: a wonderful complex piece of machinery is whirring away beautifully and then someone drops in a big ugly spanner, the machine coughs and splutters, shards of metal start to fly all around, steam and smoke burst forth and then the whole thing grinds to a halt.

I also love the expression 'you are a hero'. It's often said when someone saves the day and does something out of the ordinary to make sure everyone is happy.

Here's how you play the game.

- **Step One** Decide on a customer service issue. For this example we are going to imagine you work for a company that sells tickets with hospitality for sporting events and you are selling packages for the Wimbledon semi-final.
- **Step Two** Plan the customer service experience and as a team write down every step on an A6 card. You can go into as much detail as you like with this. So in our example something like:

1 Bulk-buy 40 tickets for a sporting event.

2 Place an advert in the sports pages of a newspaper.

3 Organise somebody to take the calls and alert them that an advert has been placed in Y paper for Z day.

4 Take incoming calls.

5 Explain the package to a potential customer.

6 Customer buys the tickets.

7 Credit card transaction takes place.

8 Tickets posted to the customer.

That's a very simple way to look at the process. You may wish to go into a lot more detail. Once all the cards are completed and all the links are made by attaching wool. Now the fun can really begin.

Split into two teams and have (depending on numbers) one or two referees. Toss a coin to decide which team will be the Spanners and which team will be the Heroes. Each member of the Spanner team gets one spanner card and each member of the Hero team gets two hero cards.

The Spanners then need to decide where they are going to place 'a spanner in the works'. The idea is to create a situation that has the maximum negative effect from the customer's perspective. Then they play a spanner card and place it on the part of the process they want to affect and explain what their spanner means. The referee decides whether it is a fair and realistic problem to have.

So in our example we may decide that a spanner in the works would be not to alert those appointed to receive the calls about the advert and the day it is due to appear. The Spanners explain that this would cause confusion because people would be calling in larger numbers than normal, they wouldn't know what advert the customer was responding to and huge amounts of stress and negativity would be caused which in turn would be passed on to the customer. A real result for the Spanners.

The Heroes then have to decide what they are going to do with their two cards.

The first fixes the problem, the second enhances the system.

So they may choose to use their first card to fix the problem by focusing on a flexible attitude with the staff and by having several staff multi-skilled so they can take over the phones should they be caught out during unexpected busy periods. The second card has to enhance a part of the current system. They may choose to enhance the piece they are working on, making it bulletproof so it doesn't happen again. Or they may choose to enhance another area such as putting in two customer service calls, one to ensure they have the tickets soon after the expected arrival date and one the day after the event to see how it went and to tell them about other events they may be interested in.

Once each team has played their cards, decide which items are going to be made a real part of your organisation's ways of working.

Remember – the aim of the game is for the **Heroes to win**!

Wee Wow

People tend to play games the same way they play life. A good facilitator will observe how people react and use this to create discussion at the end of each live issue.

51

Using 5 Star Service as a training resource

The first edition of *5 Star Service* has been used by many people as a training resource as much as an inspirational book. Almost every day we receive emails and calls from people who have used or are using *5 Star Service* with their staff.

Of course in a perfect world you would pick up the phone, call me and I'd be happy to come and present the ideas to you, but the reality is I can't be in two places at the same time (yet).

Ideas such as the Service Star™, RADAR thinking™, Wee Wows and the Spanners and Heroes game have become regular training activities for hundreds of organisations. I want more people to use *5 Star Service* as a training aid so this chapter is designed to give you a framework to implement the ideas into your organisation.

To get the best out of this chapter and to make it fair for everyone I suggest we negotiate a learning contract. I'll tell you what I'll do and then I'll suggest what you should do. If we both agree then the resources are yours.

Here's what I'll do:

I will give you a suggested 90-day (13-week) framework to suit your organisation.

I will offer you support materials to help you do this via the *Five Star Service* pages of www.michaelheppell.com.

I will also supply you with an audio introduction to several of the ideas to play to your team at training sessions.

And here's the best bit. As you've bought this book, to say thank you I won't charge you anything for everything listed above.

Here's what I would like you to do:

To agree to use the resource materials available only from my website.

I'd also like you to agree not to photocopy any part of this book. If you want more, please buy them (they are great value).

You will credit the source for any of the materials you use. 'This is an idea from Michael Heppell's book, *5 Star Service*' would be perfect!

You won't charge anyone for using these materials. If you are unsure because you are a sub-contractor, trainer, coach or presenter then contact Michael Heppell Ltd (details are at the back of this book) and ask. We normally say yes when we are asked and we generally sue when we aren't.

I've created five programmes that cover:

Five star in the office

Five star for retail

Five star for hotels and restaurants

Five star for the public sector

Five star for education

If you don't fit with any of those then take a peep at the content of each and use it as an inspiration to create your own.

When your programme works send it through to me and I'll share your success by posting it on the *5 Star Service* pages on www.michaelheppell.com.

Here are the suggested plans with a few thoughts thrown in.

Five star in the office

Week 1	**The Service Star™** Every programme starts here, it's how you know where you are and how you'll know when you're getting better!
Week 2	**Wee Wows** An easy concept that gives you some quick wins.
Week 3	**Heads up!** Once you have explored the concept be sure to get the buy-in from everyone involved. Schedule a review in a month.
Week 4	**Ring the bell** If you are the leaders it is essential that you encourage people to ring the bell over the next seven days to create a success habit.
Week 5	**Super scripts and it's not what you say** This is very powerful for telephone service, dealing with internal customers and suppliers.
Week 6	**Systemise routines – personalise exceptions** Ensure you have a few examples up your sleeve to get the discussion started.
Week 7	**Empowering staff** You need to know that what you agree to in this session can be implemented immediately. So it's good to know your limitations and breadth of flexibility at this point too.
Week 8	**Complaints – a chance to shine** Be watchful during this session to keep it upbeat and avoid any kind of blame culture.
Week 9	**What's in a name?** Have a few ideas that you have worked out here. If you belong to a large organisation it can be fun to have a name quiz featuring pictures of people from other departments too.
Week 10	**RADAR thinking™** As soon as you have completed this session take massive action to implement your ideas. Remember to schedule your review time in too.

Week 11 **Telephone service and advanced telephone service** Everyone 'thinks' they are getting the basics right but it's worthwhile reviewing them before you move on to the advanced techniques.

Week 12 **Top three referability habits** An easy one this week but don't be fooled by it. These three techniques need a good five minutes each to explore how you can do each one better.

Week 13 **Spanners and Heroes** Lots of fun, chaos and learning to be had on this your final week. This exercise can take up to one hour so, if you can, schedule some extra time.

Five star for retail

It's a tough time in retail so you have to be better than ever.

As customers demand more, you have an obligation to make every transaction magical.

Week 1 **The Service Star™** Every programme starts here. It's how you know where you are and how you'll know when you're getting better!

Week 2 **Wee Wows** An easy concept that gives you some quick wins.

Week 3 **Heads up!** Once you have explored the concept be sure to get the buy-in from everyone involved. Schedule a review in a month.

Week 4 **The emotional bank account** Link this right up to the point where you ask a customer to pay, and explore how happy they would be to hand their money over if their 'account' with you is already overdrawn.

Week 5 **One chance to make a first impression** It can be easy to point out some of the things that are wrong with your team here but instead ask your

	team to point out who does each of the points well.
Week 6	**Empowering staff and secret shopper** If you are going to use the secret shopping techniques, then now would be a good time to introduce them. When it comes to empowering staff you need to know that what you agree to in this session can be implemented immediately. So it's good to know your limitations and breadth of flexibility at this point too.
Week 7	**Complaints – a chance to shine!** Be watchful during this session to keep it upbeat and avoid any kind of blame culture.
Week 8	**Big Buyer is watching you** For this session it is very powerful to have a personal example or two in mind so you can get the ball rolling. No one likes to think they are guilty of the challenges in this area so play it safe and allow lots of discussion.
Week 9	**RADAR thinking™** As soon as you have completed this session take massive action to implement your ideas. Remember to schedule your review time in too.
Week 10	**It's your best friend – the awkward customer** This should be one of your most fun sessions with a serious message. It may be fun to hear some examples from your team of what they consider an awkward customer to be like.
Week 11	**99 per cent of people are good** I know you are probably tied by company rules and regulations here, but the idea is not to change the rules it's to reframe how you handle them.
Week 12	**Sell me a solution** Get this right and you'll see confidence and profits soar! The idea is very simple, the application is not. Take time to role-play ideas and perhaps incorporate super scripts.

Week 13 **Spanners and Heroes** Lots of fun, chaos and learning to be had on this your final week. This exercise can take up to one hour so, if you can, schedule some extra time.

Five star for hotels and restaurants

Choosing just 13 sessions for you is a real challenge as you'll want to be brilliant at everything in this book. However, you have to start somewhere and this programme gives you a wide range of skills and some timely reminders of what you should be doing.

Week 1 **The Service Star™** Every programme starts here. It's how you know where you are and how you'll know when you're getting better!

Week 2 **One chance to make a first impression** It can be easy to point out some of the things that are wrong with your team here but instead, ask your team to point out who does each of the points well.

Week 3 **Heads up!** Once you have explored the concept be sure to get the buy-in from everyone involved. Schedule a review in a month.

Week 4 **Wee Wows** An easy concept that gives you some quick wins.

Week 5 **Beware the silent customer** This week you can really play up the idea of outstanding five star service and creating advocates of your hotel or restaurant.

Week 6 **RADAR thinking™** As soon as you have completed this session take massive action to implement your ideas. Remember to schedule your review time in too.

Week 7 **Customer magic moments** Read the chapter about Captain Denny Flanagan and how he treats every customer as if they are flying for

	the first time. This is a magical way to start your session.
Week 8	**The blind spot** I think you will be surprised at how many 'blind spots' you have when you start to discuss this issue. If someone shares an example, be sure not to ask, 'Why didn't you mention it?' and make sure you do say, 'Thank you.'
Week 9	**The distraction of dirt** Not an easy one but a vital one. There will be plenty of people who would like to hand over responsibility in this area to someone else. The secret is to make sure everyone feels like they can make a positive difference.
Week 10	**What's in a name?** Have a few ideas that you have worked out here. If you belong to a large organisation it can be fun to have a name quiz featuring pictures of people from other departments or shifts too.
Week 11	**Special requirements** The main purpose of this session is to change thinking from one of 'how inconvenient' to one of 'wow, our chance to shine'. Gather a few real-life examples first and use this session as a chance to educate.
Week 12	**It's not what you say** It's week 12 now and you should have the confidence to role-play a little and test out some of the ideas more thoroughly.
Week 13	**Spanners and Heroes** Lots of fun, chaos and learning to be had on this your final week. This exercise can take up to one hour so, if you can, schedule some extra time.

Five star for the public sector

Having worked with dozens of public-sector organisations including the NHS and regional police forces, I don't accept that it's a more difficult environment in which to create five star service. You just have to approach it in a different way. The programme I've outlined here should excite anyone who is committed to public service.

Week 1	**The Service Star™** Every programme starts here. It's how you know where you are and how you'll know when you're getting better!
Week 2	**Wee Wows** An easy concept that gives you some quick wins.
Week 3	**Heads up!** Once you have explored the concept be sure to get the buy-in from everyone involved. Schedule a review in a month.
Week 4	**The emotional bank account** Public sector differs here as customers feel they have already 'paid' for their service. This means you often start from a place of withdrawal. Explore what that means.
Week 5	**Systemise routines – personalise exceptions** Ensure you have a few examples up your sleeve to get the discussion started.
Week 6	**RADAR thinking™** As soon as you have completed this session take massive action to implement your ideas. Remember to schedule your review time in too.
Week 7	**What's in a smile?** This should provide you with lots of quick wins. I know you want to smile, I know you do smile, but public perception is often different. This is your chance to change that.
Week 8	**Speed it up!** Our public sector has a reputation for taking weeks when a day would do. I know you can't change everything and other people may slow you down, so focus on what you *can* do.

Week 9	**Complaints – a chance to shine!** Be watchful during this session to keep it upbeat and avoid any kind of blame culture.
Week 10	**It's not what you say** It's week 10 now and you should have the confidence to role-play a little and test out some of the ideas more thoroughly.
Week 11	**Service PR** At this point in your training what are you proud of? Contact your press department with some good news stories about your five star service. Be proud!
Week 12	**Service values** This could be your most challenging week as you may not always feel you can change such a huge organisation. Again focus on what you can do.
Week 13	**Spanners and Heroes** Lots of fun, chaos and learning to be had on this your final week. This exercise can take up to one hour so, if you can, schedule some extra time.

Five star for education

Some of my most exciting work is with education – it's also some of my most challenging. Education and its delivery is going through times of great change and your customers – students, parents, employers, government, etc have very high demands.

Week 1	**The Service Star™** Every programme starts here. It's how you know where you are and how you'll know when you're getting better!
Week 2	**Wee Wows** An easy concept that gives you some quick wins.
Week 3	**Heads up!** Once you have explored the concept be sure to get the buy-in from everyone involved. Schedule a review in a month.
Week 4	**The emotional bank account** Education differs here as customers could feel they have already

	'paid' for their service. This means you often start from a place of withdrawal. Explore what that means.
Week 5	**Systemise routines – personalise exceptions** Ensure you have a few examples up your sleeve to get the discussion started.
Week 6	**RADAR thinking™** As soon as you have completed this session take massive action to implement your ideas. Remember to schedule your review time in too.
Week 7	**One chance to make a first impression and the distraction of dirt** It can be easy to point out some of the things that are wrong with your team here but instead ask your team to point out who does each of the points well. I've incorporated the distraction of dirt too, not an easy one to address but a vital one. There will be plenty of people who would like to hand over responsibility in this area to someone else. The secret is to make sure everyone feels like they can make a positive difference.
Week 8	**What's in a smile?** This should provide you with lots of quick wins. I know you want to smile, I know you do smile, but public perception is often different. This is your chance to change that.
Week 9	**Making the mundane marvellous** Especially with young people. Education can seem like a drag (we all know they'll think differently in a few years) so it's important to make each part of the learning process, from enrolment to graduation, interesting, stimulating and enjoyable.
Week 10	**Prepare for and relish competition** Education is becoming more and more competitive with learners being offered endless choice. How can you use this to your advantage? It's everyone's responsibility to be active in this.

Week 11	**It's not what you say** It's week 11 now and you should have the confidence to role-play a little and test out some of the ideas more thoroughly.
Week 12	**Service values** This could be your most challenging week as you may not always feel that you can change such large organisations. Again focus on what you can do.
Week 13	**Spanners and Heroes** Lots of fun, chaos and learning to be had on this your final week. This exercise can take up to one hour so, if you can, schedule some extra time.

And finally ...

So you've got a whole bunch of tools, anecdotes, techniques and strategies. How are you going to use them? Or if you've already started, what have you found?

As I said right at the start of this book, the secret isn't in the knowing, it's in the doing. Ideas are just ideas, doing great ideas makes the difference. Edison didn't just think about the light bulb, he built one, then once he had built it he opened a company to make and distribute the lights. Some would argue that his greatest success wasn't the invention of the incandescent electric light bulb, it was giving people light.

The same can be said of great service. It's not really about the service techniques, *it's how it makes people feel*. That's what's really special – and you can do that. You can make people feel better about you, your organisation, even themselves! Isn't that great?

But it doesn't happen by accident and it certainly doesn't happen just by reading this book. Think of this book like the manual for a new car. If you've bought a new car recently you'll notice you tend to get two books – a quick guide and a detailed manual. The quick guide is designed to get you started and to use as an instant reference. You can use this book like that by taking the basic ideas you've learned and applying them. Or you could go a bit deeper and think of this book as the more detailed manual. The difference will be your application, and also the difference in the payback and rewards is vast.

Quick-guide benefits and results:

- Hop in and out – you're busy so just dive in and test the bits that seem most relevant to you.
- Look for some quick fixes (there are some good ones) and see an immediate improvement in your customer service skills.
- Share the relevant bits with your colleagues. They can take your highlights and apply them too, getting you on the same page quickly.

The 'detailed manual' approach:

- Create a cultural change in your organisation.
- Use *5 Star Service* as a manual for your ways of working.

- Take a section each week and spend 15 minutes as a team looking at how you can apply the ideas. Measure the impact and results and choose which work best for you.
- Keep a journal of the ideas you've used and the impact they have. Share this information and write your own ways to make the ideas work best for you and your organisation.

Whichever way you choose, I hope that by reading *5 Star Service* you have become as passionate as I am about customer service.

And passionate is not an accidental choice of word.

When you're **passionate** about something you'll **overcome any obstacles** to make it happen. You'll **find a way** when others say it can't be done. You'll **get creative** when all around you seems staid.

Passion will give you energy when you should have had enough. Passion for people is what's needed to make the ideas in this book really work. If you haven't got it yet – read it again!

From the Service Star™ at the beginning to the realisation that your competition is in fact everyone, I've given you tools and actions to put in place to *get results*. I know what it's like working in or running an organisation these days. It's tough – very tough. But it's now more than ever that you need to differentiate yourself from the competition, and I believe the best way to do it is through your amazing five star service.

It's not easy, but then it's not really hard either. It's not quick, but then it isn't slow. It's certainly cheap in material costs but can be high in emotional ones. Whichever way you look at it, creating five star magic moments for the people you meet and connect with every day has got to be one of the most rewarding actions you can take.

I hope to have the opportunity to experience your five star service in my life soon, as I hope you'll have the chance to experience mine in yours.

Enjoyed *5 Star Service*?

Free Bonus Chapters

If you would like to receive the bonus chapters mentioned in this book simply send an email to info@michaelheppell.com with the heading '5 Star Bonus Chapters' and we'll send them directly to you.

Plus you can download 5 Star resources by visiting The Michael Heppell website: www.michaelheppell.com.

There are several pages of extra goodies, downloads and resources on the *5 Star Service* pages of www.michaelheppell.com too.

Book Michael Heppell for your next event

Would you like Michael to speak to your organisation?

Michael has been described as one of the top speakers on the planet! His keynote presentations and special events have been enjoyed by audiences around the world.

If you would like to know more about booking Michael to speak to your organisation then contact the Michael Heppell Ltd team.

Tel: UK 08456 733 336
International +44 1434 688 555
Email: info@michaelheppell.com
Website: www.michaelheppell.com

Contact Michael Heppell

Michael can be contacted directly by emailing:

info@michaelheppell.com or by calling (from the UK) 08456 733 336 (or international) +44 1434 688 555

www.michaelheppell.com

For television enquiries contact:

Michael Foster at MF Management
Tel: 0203 291 2929
email: mfmall@mfmanagement.com

What real readers said about *How to be Brilliant*

'Not only have I read it, I'm re-reading and re-reading, and re-doing! Love the book. Clear, straightforward, love the 'no-bull' and absence of jargon!'

Josephine Cropper
Healthy Horizons Psychotherapy

'Just to let you know I've read your book and will be reading it again and recommending it to my friends and colleagues. One of the best for a long time and believe me I've read them all!'

Diane Inverarity
Sales and Education
Clinique, Estée Lauder Companies

'I have so far given this book to 90 call centre staff, 32 leads and I have recommended it to my mentor ring which has people from 10 different countries on it.

If you like Stephen Covey then you will love Michael Heppell – he is in my opinion the UK's version of Steven Covey.'

Karen Young
Microsoft (EMEA Regional Manager for MVP and Communities)

'After a company restructure I found myself made redundant from a high profile job. In preparation for interviews, a friend recommended your book. At 3 o'clock this morning I turned the bedside light off after reading half the book. I'm just now reading the rest and emailing 20 or so friends to recommend they buy it too. Simple message ... it's Brilliant!'

MW

'I often pick up your book and read a few pages just to remind myself where I want to be.'

GG

'I have read *How to be Brilliant* and I will recommend it to anyone in search of a brilliant book on personal development.'

HY

'This is the best written and most stimulating book of its type that I've ever read! Thanks a lot for your wise advice.'

RT

'I've just finished reading your book *How to be Brilliant.* It's absolutely refreshing, inspiring and most of all BRILLIANT.'

UU

'I got your book today. I have had things to do, places to go and people to see – but I still managed to read at least half your book, do the exercises, make notes and it's only 6pm! I can't put it down! I feel fired up. It is so right for me just now, and I know other people who will benefit from your easy, lively, entertaining and direct brilliant writing style – it is top of my list for "present of the year".'

SW

'I recently saw my niece (we don't see each other very often) – she was complaining about her job – boring/not well paid/no prospects – she was managing a (not very busy) shop. I just happened to have to hand your book – which I told her about and pointed out a few relevant areas – and said "well worth it if you've got a spare £10". I bumped into her again about an hour ago – the first thing she said was "I got that book ... and I've got a new job, got a company car and the pay is great!" I couldn't help but say "Brilliant!".'

GI

'The book lives up to its title. Definitely the most brilliant book I have ever read.'

TW

'I just bought your book yesterday – hardly able to put it down. The physical layout of your book and your very practical and warm writing style have both made it an exceptional read so far, great job, really finding it has impact for me!'

NH

'I read your book *How to be Brilliant* and got started immediately. Brilliant results at once. Thank you for the inspiration and the tips to be brilliant.'

CC

'This is an incredible book and so inspirational and practical.'

JV

'Your book is fantastic, so clear and straightforward. I have referred many of my friends and colleagues to it already.'

AD

'I love the book so much I have bought a further three copies and given them to friends and all for different reasons. I think the book is straightforward and a brilliant tool box.'

NM

'I love your book. Very inspirational and I'm taking massive action and seeing the differences immediately. This is a book I will buy and give to friends.'

TP